Determination of Elements in Natural-Water, Biota, Sediment, and Soil Samples Using Collision/Reaction Cell Inductively Coupled Plasma–Mass Spectrometry

By John R. Garbarino, Leslie K. Kanagy, and Mark E. Cree

Chapter 1
Section B, Methods of the National Water Quality Laboratory
Book 5, Laboratory Analysis

Techniques and Methods 5–B1

U.S. Department of the Interior
U.S. Geological Survey

U.S. Department of the Interior
Gale A. Norton, Secretary

U.S. Geological Survey
P. Patrick Leahy, Acting Director

U.S. Geological Survey, Reston, Virginia: 2006

For sale by U.S. Geological Survey, Information Services
Box 25286, Denver Federal Center
Denver, CO 80225

For more information about the USGS and its products:
Telephone: 1-888-ASK-USGS
World Wide Web: http://www.usgs.gov/

Suggested citation:
Garbarino, J.R., Kanagy, L.K., and Cree, M.E., 2006, Determination of elements in natural-water, biota, sediment, and soil samples using collision/reaction cell inductively coupled plasma–mass spectrometry: U.S. Geological Survey Techniques and Methods, book 5, sec. B, chap. 1, 88 p.

Contents

Figures

Appendix Figures

Tables

CONVERSION FACTORS, ABBREVIATED WATER-QUALITY UNITS, ADDITIONAL ABBREVIATIONS, AND DEFINITIONS

Multiply	By	To obtain
centimeter (cm)	3.94×10^{-1}	inch
millimeter (mm)	3.94×10^{-2}	inch
micrometer (μm)	3.94×10^{-5}	inch
Mass		
gram (g)	3.53×10^{-2}	ounce, avoirdupois
milligram (mg)	3.53×10^{-5}	ounce, avoirdupois
microgram (μg)	3.53×10^{-8}	ounce, avoirdupois
Volume		
liter (L)	2.64×10^{-1}	gallon
milliliter (mL)	2.64×10^{-4}	gallon
microliter (μL)	2.64×10^{-7}	gallon

Degree Celsius (°C) may be converted to degree Fahrenheit (°F) by using the following equation:

$$°F = 9/5 \ (°C) + 32$$

Abbreviated water-quality units used in this report:

μg/g	micrograms per gram
μg/L	micrograms per liter
μL/min	microliters per minute
μS/cm	microsiemens per centimeter at 25°C
mg/g	milligrams per gram
mg/L	milligrams per liter
mL/min	milliliters per minute

Other abbreviations used in this report:

ASTM	American Society for Testing and Materials
cICP–MS	collision/reaction cell inductively coupled plasma–mass spectrometry
GF–AAS	graphite furnace–atomic absorption spectrometry
HPLC	high-performance liquid chromatography
ICP–MS	inductively coupled plasma–mass spectrometry
ICP–AES	inductively coupled plasma–atomic emission spectrometry
LT–MDL	long-term method detection level
MDL(s)	method detection limit(s)
m/z	mass-to-charge ratio
NIST	National Institute of Standards and Technology

Other abbreviations used in this report (Continued):

NWQL	National Water Quality Laboratory
OctP	octapole
P/A	pulse/analog factor
QP	quadrupole
SRM	standard reference material
SRWS(s)	U.S. Geological Survey standard reference water sample(s)
USGS	U.S. Geological Survey
% Bias	percent error between the expected concentration and the experimental mean concentration
% RSD	percent relative standard deviation
≥	greater than or equal to
%	percent
±	plus or minus

Concentrations of chemical constituents in water are given either in milligrams per liter (mg/L) or micrograms per liter (µg/L). Concentrations of chemical constituents in solid digestates are given either in milligrams per gram (mg/g) or micrograms per gram (µg/g), dry weight.

DETERMINATION OF ELEMENTS IN NATURAL-WATER, BIOTA, SEDIMENT, AND SOIL SAMPLES USING COLLISION/REACTION CELL INDUCTIVELY COUPLED PLASMA–MASS SPECTROMETRY

By John R. Garbarino, Leslie K. Kanagy, and Mark E. Cree

Abstract

A new analytical method for the determination of elements in filtered aqueous matrices using inductively coupled plasma–mass spectrometry (ICP–MS) has been implemented at the U.S. Geological Survey National Water Quality Laboratory that uses collision/reaction cell technology to reduce molecular ion interferences. The updated method can be used to determine elements in filtered natural-water and other filtered aqueous matrices, including whole-water, biota, sediment, and soil digestates. Helium or hydrogen is used as the collision or reaction gas, respectively, to eliminate or substantially reduce interferences commonly resulting from sample-matrix composition. Helium is used for molecular ion interferences associated with the determination of As, Co, Cr, Cu, K, Mg, Na, Ni, V, W and Zn, whereas hydrogen is used for Ca, Fe, Se, and Si. Other elements that are not affected by molecular ion interference also can be determined simply by not introducing a collision/reaction gas into the cell. Analysis time is increased by about a factor of 2 over the previous method because of the additional data acquisition time in the hydrogen and helium modes.

Method detection limits for As, Ca, Co, Cr, Cu, Fe, K, Mg, Na, Ni, Se, Si (as SiO_2), V, W, and Zn, all of which use a collision/reaction gas, are 0.06 microgram per liter (µg/L) As, 0.04 milligram per liter (mg/L) Ca, 0.02 µg/L Co, 0.02 µg/L Cr, 0.04 µg/L Cu, 1 µg/L Fe, 0.007 mg/L K, 0.009 mg/L Mg, 0.09 mg/L Na, 0.05 µg/L Ni, 0.04 µg/L Se, 0.03 mg/L SiO_2, 0.05 µg/L V, 0.03 µg/L W, and 0.04 µg/L Zn. Most method detection limits are lower or relatively unchanged compared to earlier methods except for Co, K, Mg, Ni, SiO_2, and Tl, which are less than a factor of 2 higher.

Percentage bias for samples spiked at about one-third and two-thirds of the concentration of the highest calibration standard ranged from –8.1 to 7.9 percent for reagent water, –14 to 21 percent for surface water, and –16 to 16 percent for ground water. The percentage bias for reagent water spiked at trace-element concentrations of 0.5 to 3 µg/L averaged 4.4 percent with a range of –6 to 16 percent, whereas the average percentage bias for Ca, K, Mg, Na, and SiO_2 was 1.4 percent with a range of –4 to 10 percent for spikes of 0.5 to 3 mg/L. Elemental results for aqueous standard reference materials compared closely to the certified concentrations; all elements were within 1.5 F-pseudosigma of the most probable concentration. In addition, results from 25 filtered natural-water samples and 25 unfiltered natural-water digestates were compared with results from previously used methods using linear regression analysis. Slopes from the regression analyses averaged 0.98 and ranged from 0.87 to 1.29 for filtered natural-water samples; for unfiltered natural-water digestates, the average slope was 1.0 and ranged from 0.83 to 1.22. Tests showed that accurate measurements can be made for samples having specific conductance less than 7,500 microsiemens per centimeter (µS/cm) without dilution; earlier ICP–MS methods required dilution for samples with specific conductance greater than 2,500 µS/cm.

Introduction

Inductively coupled plasma–mass spectrometry (ICP–MS) has been used to determine elements in natural-water samples at the U.S. Geological Survey National Water Quality Laboratory since the early 1990s (Faires, 1993; Garbarino and Taylor, 1994; Garbarino and Struzeski, 1998; Garbarino, 1999; Garbarino, 2000; Garbarino and others, 2002). Methods have been updated over the years to document changes in data acquisition modes, to expand the number of elements determined, and to expand the types of matrices analyzed. This new ICP–MS method was implemented to use recent advances in collision/reaction cell technology that reduce molecular ion interference resulting in improved accuracy and in expanded scope.

A molecular ion interferes when its mass cannot be resolved from the mass of the element of interest because of limitations on the resolution of the quadrupole mass spectrometer. The most prominent molecular-ion interferences are associated with oxide and hydroxide species arising from matrix elements in the sample (for details, see Horlick and Montaser, 1998). One example of such interference is $^{35}Cl^{16}O^+$ on the determination of $^{51}V^+$. Other molecular-ion interferences, such as $^{40}Ar^{35}Cl^+$ on $^{75}As^+$ and $^{40}Ar^{23}Na^+$ on $^{63}Cu^+$, are associated with elements in the plasma and sample matrix. Before the introduction of collision/reaction cell technology, such interference corrections were commonly made by measuring the signal at an associated interference-free isotope and subtracting the fraction corresponding to the interference from the signal of the quantifying isotope. Although this approach was fairly successful, analyte concentrations in the micrograms-per-liter range likely were to become more negatively biased with increasing concentrations of matrix interferents. Alternatively, the formation of oxide species could be reduced by using cold plasma conditions or by using a different elemental isotope whenever possible. However, either approach usually compromised the sensitivity of the analysis.

Using a collision/reaction cell system with an ICP–MS reduces molecular ion interferences through either collisional or reactional processes. Helium is used as the cell gas to promote collisional dissociation and energy discrimination of molecular ions that cause interference. A molecular ion at the same nominal mass-to-charge ratio (m/z) and kinetic energy as an analyte ion experiences more collisions because of its larger cross-sectional area. As a result of these collisions, the kinetic energy of the molecular ion is reduced, and it is rejected by the cell, whereas the analyte ion is transmitted through the quadrupole to the detector. When hydrogen is used as the cell gas, simple reactions with argon-based molecular ions take place through charge-, proton-, or atom-transfer processes. Analyte ions generally do not react with hydrogen; therefore, there is no loss in analyte signal. Reaction processes can produce new molecular ions, principally hydrides of the original molecular-ion interference. However, such species have low kinetic energies and usually do not interfere.

Purpose and Scope

This report describes a new method for the determination of As, Co, Cr, Cu, Ni, Se, V, and Zn in filtered water, unfiltered water digestates, biota digestates, sediment digestates, and soil digestates using collision/reaction cell inductively coupled plasma–mass spectrometry (cICP–MS). In addition, other elements, such as Ca, Fe, K, Mg, Na, Si, and W that were not determined routinely by earlier ICP–MS methods, have been added to the method. Furthermore, the method can be used to determine arsenic species using high-performance liquid chromatography (HPLC) for separation and cICP–MS for detection. Elements that were determined in earlier ICP–MS methods and that are not affected by molecular ion interference also are determined by cICP–MS by simply not

introducing hydrogen or helium into the gas cell. New laboratory codes and parameter codes are not needed for Ag, Al, B, Ba, Be, Cd, Li, Mn, Mo, Pb, Sb, Sr, Tl, and U because the plasma conditions, acquisition characteristics, and analytical performance for such elements are not drastically different from earlier ICP–MS methods that did not use reaction/collision cell technology.

Information is provided in this report to:

- Establish the method detection limits (MDLs) for all elements determined using the cICP–MS method.

- Evaluate the capability of collision/reaction cell technology to reduce molecular ion interferences on the determination of Ag, Al, As, Ca, Cd, Co, Cr, Cu, Fe, K, Mg, Mn, Na, Ni, Se, Si, V, W, and Zn.

- Review sample dilution guidelines using spiked field samples having specific conductance $\geq 2,000$ μS/cm.

- Determine the analytical variability of measurements in laboratory reagent water as a function of analyte concentration.

- Determine the bias and variability of the cICP–MS method using results from standard reference materials.

- Validate the cICP–MS method for the determination of elements in ground-water, surface-water, and laboratory reagent-water matrices.

- Compare the bias and variability of cICP–MS with results from other analytical methods for 25 filtered natural-water samples and 25 whole-water digestates.

The method described was developed by the U.S. Geological Survey (USGS) for use at the National Water Quality Laboratory (NWQL). This method replaces ICP–MS method I-2477-92 for the determination of dissolved As, Co, Cr, Cu, Ni, Se, V, and Zn (Faires, 1993; Garbarino, 1999), methods I-4471-97 and I-4472-97 for the determination of whole-water recoverable As, Co, Cr, Cu, Ni, Se, V, and Zn (Garbarino, 2000; Garbarino and Struzeski, 1998), methods I-1190-02, I-2191-02, I-2193-02, and I-2192-02 for arsenic speciation (Garbarino and others, 2002). The new method was implemented at the NWQL in October 2005.

Analytical Method

New laboratory codes, parameter codes, and method codes for collision/reaction cell inductively coupled plasma–mass spectrometry (cICP–MS) are listed in table 1 for As (speciated and unspeciated), Co, Cr, Cu, Ni, Se, V, and Zn in different sample matrices. New codes also were established for Ca, Mg, Fe, K, Na, Si (as silica), and W for filtered and unfiltered water matrices. Laboratory codes, parameter codes, and method codes for Ag, Al, B, Ba, Be, Cd, Li, Mn, Mo, Pb,

Sb, Sr, Tl, and U, elements not affected by molecular ion interference, remain unchanged for each type of sample matrix.

Application and Method Detection Limits

The method described in this report can be used to determine a wide range of elements in filtered water, unfiltered water digestates, biological digestates, sediment digestates, and soil digestates. The collision/reaction gas is used when determining As (speciated and unspeciated), Ca, Co, Cr, Cu, Fe, K, Mg, Na, Ni, Se, Si, V, W, and Zn. Method detection limits are 0.06 microgram per liter (μg/L) As, 0.04 milligram per liter (mg/L) Ca, 0.02 μg/L Co, 0.02 μg/L Cr, 0.04 μg/L Cu, 1 μg/L Fe, 0.007 mg/L K, 0.009 mg/L Mg, 0.09 mg/L Na, 0.05 μg/L Ni, 0.04 μg/L Se, 0.03 mg/L SiO_2, 0.05 μg/L V, 0.03 μg/L W, and 0.04 μg/L Zn (see table 2). Most method detection limits are lower or relatively unchanged compared to earlier methods except for Co, K, Mg, Ni, SiO_2, and Tl, which are less than a factor of 2 higher. Other elements (Ag, Al, B, Ba, Be, Cd, Li, Mn, Mo, Pb, Sb, Sr, Tl, and U) that have been determined using previous ICP–MS methods can be

Table 1. Codes for elements in water, biota, and sediment and for arsenic species in water samples determined by collision/reaction cell inductively coupled plasma-mass spectrometry.

[cICP-MS, collision/reaction cell inductively coupled plasma-mass spectrometry; I-xxxx-xx, method number; Lab code, laboratory code; P code, parameter code and method code; μg/L, micrograms per liter; mg/L, milligrams per liter; μg/g, micrograms per gram dry weight]

Element	Lab code	P code	Lab code	P code	Species	Lab code	P code
	Elements, water, filtered, cICP–MS, I-2020-05		Elements, water, unfiltered, cICP–MS, I-4020-05			Arsenic species, water, filtered, field separation, cICP–MS, I-2197-05	
As, μg/L	3122	01000 H	3123	01002 H	Arsenite [As(III)], H_3AsO_3	3140	62452 E
Ca, mg/L	2916	00915 I	2917	00916 D	Arsenate [As(V)], $H_2AsO_4^-$	3140	62453 E
Co, μg/L	3124	01035 I	3125	01037 I			
Cr, μg/L	3126	01030 J	3127	01034 I		Arsenic species, water, filtered, laboratory separation, malonate/acetate mobile phase, cICP–MS, I-2196-05	
Cu, μg/L	3128	01040 I	3129	01042 I			
Fe, μg/L	2974	01046 I	2975	01045 D			
K, mg/L	3014	00935 E	3015	00937 D	Arsenite [As(III)], H_3AsO_3	3141	62452F
Mg, mg/L	2984	00925 I	2985	00927 D	Arsenate [As(V)], $H_2AsO_4^-$	3141	62453 F
Na, mg/L	3048	00930 I	3049	00929 D			
Ni, μg/L	3130	01065 I	3131	01067 J		Arsenic species, water, filtered, laboratory separation, phosphate mobile phase with arsine generation, cICP–MS, I-2195-05	
Se, μg/L	3132	01145 G	3133	01147 H			
SiO_2, mg/L	3046	00955 J	3047	00956 B			
V, μg/L	3134	01085 G	3135	01087 D	Arsenite [As(III)], H_3AsO_3	3142	62452 G
W, μg/L	3136	01155 A	3137	01154 A	Arsenate [As(V)], $H_2AsO_4^-$	3142	62453 G
Zn, μg/L	3138	01090 I	3139	01092 E	Monomethylarsonate (MMA), $(CH_3)HAsO_3^-$	3142	62454 E
	Elements, biota, cICP–MS, I-9020-05		Elements, sediment or soil, recoverabe, cICP–MS, I-5020-05		Dimethylarsinate (DMA), $(CH_3)_2HAsO_2$	3142	62455 E
As, μg/g	6055	49247 C	3144	64847 A		Arsenic species, water, filtered, laboratory separation, nitric acid mobile phase, cICP–MS, I-2193-05	
Co, μg/g	6056	49250 D	3145	01038 E			
Cr, μg/g	6057	49240 C	3146	01029 E			
Cu, μg/g	6058	49241 C	3147	01043 E	Arsenite [As(III)], H_3AsO_3	3143	62452 H
Ni, μg/g	6059	49253 D	3148	01068 D	Arsenate [As(V)], $H_2AsO_4^-$	3143	62453 H
Se, μg/g	6060	49254 C	3149	64848 A	Monomethylarsonate (MMA), $(CH_3)HAsO_3^-$	3143	62454 F
V, μg/g	6061	49465 C	3150	64849 A	Dimethylarsinate (DMA), $(CH_3)_2HAsO_2$	3143	62455 F
Zn, μg/g	6062	49245 C	3151	01093 C			

determined with similar performance using this method. The typical linear calibration range extends from the method detection limit (MDL) to 100 µg/L for each trace element or species. Selected elements like Al, Ba, Mn, Sr, and Zn are calibrated up to 2,500 µg/L. Elements such as Ca, K, Mg, and Na normally are calibrated up to 100 mg/L, whereas Si and Fe are calibrated up to 50 mg/L and 1,000 µg/L, respectively.

The method detection limits for all elements and species determined in filtered-water matrices using cICP–MS and long-term method detection levels (LT–MDL) from earlier methods are listed in table 2. These limits were calculated using guidelines of the U.S. Environmental Protection Agency (2000). Most method detection limits are lower or relatively unchanged compared to earlier methods except for Co, K, Mg, Ni, SiO_2, and Tl, which are less than a factor of 2 higher. Method detection limits for unfiltered water digestates are similar to those listed in table 2 except for vanadium and arsenic, which are about a factor of 10 lower than earlier methods. Method detection limits (in micrograms per gram, dry weight) for elements determined in biota, sediment, and soil digestates can be estimated by multiplying the limits in table 2 by the digestate volume (typically 0.100 L) and dividing it by the sample dry weight (typically 0.5 g). The linear calibration range can be extended when the analog stage of the detector is calibrated, otherwise samples that have elemental concentrations that exceed the upper calibration standard must be diluted or, in the case of arsenic speciation, a smaller sample volume can be injected. In addition, samples with specific conductance greater than 7,500 µS/cm need to be diluted prior to analysis.

Interferences

Spectral interferences associated with isobaric ions and molecular ions can affect the accuracy of elemental analysis using ICP–MS. Such interferences have been documented in the literature (Tan and Horlick, 1986; Horlick and Montaser, 1998). Molecular ion interferences evolve from the argon plasma and elements composing the sample matrix. Molecular ions that are potential isobaric interferences for the elements determined are listed in table 3. Such interferences are greatly reduced or eliminated for ICP–MS instruments that use a collision/reaction cell without using correction equations. By introducing a gas into the cell, either He or H_2 in this method, an interfering molecular ion is isolated from the analyte ion through collisional or reactional interactions, thereby improving the accuracy of the determination of selected elements.

Test solutions containing analytes and elements associated with molecular ion interferences were analyzed

Table 2. Method detection limits for elements and species determined using collision/reaction cell inductively coupled plasma–mass spectrometry.

[Analyte, concentrations are in micrograms per liter unless otherwise noted; m/z, mass-to-charge ratio; Mode, either normal (none), He (helium), or H_2 (hydrogen) cell gas; MDL, method detection limit; LT–MDL, long-term method detection level as of 2005; mg/L, milligrams per liter; cICP–MS, collision/reaction cell inductively coupled plasma–mass spectrometry; ICP–MS, inductively coupled plasma–mass spectrometry; ICP–AES, inductively coupled plasma–atomic emission spectrometry; GF–AAS, graphite furnace–atomic absorption spectrometry; SiO_2, determined as silicon (Si) but reported as silica (SiO_2); na, not applicable]

Analyte	m/z	cICP–MS		Previous method	
		Mode	MDL	LT–MDL	Technique
Ag	107	Normal	0.07	0.1	ICP–MS
Al	27	Normal	0.4	0.8	ICP–MS
As	75	He	0.06	0.1	ICP–MS
B	11	Normal	2	4	ICP–MS
Ba	137	Normal	0.06	0.1	ICP–MS
Be	9	Normal	0.02	0.03	ICP–MS
Ca, mg/L	40	H_2	0.04	0.02	ICP–AES
Cd	111	Normal	0.02	0.02	ICP–MS
Co	59	He	0.02	0.008	ICP–MS
Cr	52	He	0.02	0.4	GF–AAS
Cu	63	He	0.04	0.2	ICP–MS
Fe	56	H_2	1	3	ICP–AES
K, mg/L	39	He	0.007	0.005	ICP–AES
Li	7	Normal	0.1	0.3	ICP–MS
Mg, mg/L	24	He	0.009	0.004	ICP–AES
Mn	55	Normal	0.07	0.1	ICP–MS
Mo	95	Normal	0.06	0.2	ICP–MS
Na, mg/L	23	He	0.09	0.1	ICP–AES
Ni	60	He	0.05	0.03	ICP–MS
Pb	206, 207, 208	Normal	0.01	0.04	ICP–MS
Sb	121	Normal	0.02	0.1	ICP–MS
Se	78	H_2	0.04	0.2	ICP–MS
SiO_2, mg/L	28	H_2	0.03	0.02	ICP–AES
Sr	88	Normal	0.08	0.2	ICP–MS
Tl	205	Normal	0.03	0.02	ICP–MS
U	238	Normal	0.002	0.02	ICP–MS
V	51	He	0.05	0.07	ICP–MS
W	182	He	0.03	na	na
Zn	66	He	0.04	0.3	ICP–MS

Table 2. Method detection limits for elements and species determined using collision/reaction cell inductively coupled plasma–mass spectrometry.—Continued

[HPLC/cICP–MS; high-performance liquid chromatography/collision/reaction cell inductively coupled plasma–mass spectrometry using a phosphate mobile phase and hydride generation; Analyte, concentrations are in micrograms per liter; m/z, mass-to-charge ratio; Mode, He (helium), cell gas; MDL, estimated method detection limit based on non-chromatographic determination; LT–MDL, long-term method detection level as of 2005; HPLC/ICP–MS, high-performance liquid chromatography/inductively coupled plasma–mass spectrometry]

Analyte	m/z	HPLC/cICP–MS		Previous method	
		Mode	MDL	LT–MDL	Technique
As(III)	75	He	0.09	0.3	HPLC/ICP–MS
As(V)	75	He	0.1	0.4	HPLC/ICP–MS
MMA	75	He	0.2	0.6	HPLC/ICP–MS
DMA	75	He	0.09	0.3	HPLC/ICP–MS

to evaluate the performance of cICP–MS. The effects of increasing concentrations of chloride on the recovery of 5 µg/L arsenic and vanadium are shown in figures 1 and 2, respectively. Two different sources of chloride, calcium chloride ($CaCl_2$), and hydrochloric acid (HCl) were investigated. The data corresponding to measurements made without using the He cell gas shows an apparent increase in $^{75}As^+$ concentration as a result of $^{40}Ar^{35}Cl^+$. Previous ICP–MS methods corrected for such interference by using equations based on other measurements, whereas no correction equations are needed with the new method when He is used. The advantage of this approach is the elimination of errors resulting from measuring low analyte concentration in the presence of high interferent concentration. The average arsenic concentrations measured by the new method in the $CaCl_2$ and HCl solutions having up to 5,000 mg/L chloride were 5.38±0.09 and 5.38±0.08 µg/L, respectively. Similar results are shown for vanadium (see fig. 2). When He is used, the $^{35}Cl^{16}O^+$ ion is eliminated and does not contribute to the vanadium signal. The average vanadium concentration measured in the $CaCl_2$ solution was 5.06±0.08 µg/L, whereas the concentration in the HCl was 5.2±0.4 µg/L. The median chloride concentration for samples submitted to NWQL in 2004 was 16 mg/L; 95 percent of the samples had chloride concentration less than 370 mg/L.

The effect of increasing concentrations of carbon on the recovery of about 20 µg/L chromium is shown in figure 3. Solutions containing dissolved organic carbon from the Suwannee River (Florida) and Big Soda Lake (Nevada) were used to test the effectiveness of cICP–MS in eliminating the $^{40}Ar^{12}C^+$ interference on $^{52}Cr^+$. The dissolved organic carbon concentration ranged from 0 to about 20 mg/L; the maximum concentration tested is at least a factor of 10 greater than most samples submitted to NWQL. The results show that carbon molecular ion interference on chromium determina-

tions was eliminated for the two types of dissolved organic carbon tested. The average chromium concentrations measured in the Suwannee River and Big Soda Lake solutions were 22.8±0.2 and 22.6±0.5 µg/L, respectively.

Other potential molecular ion interferences that are listed in table 3 also were evaluated. The percentage spike recovery for each analyte in the presence of elements that are associated with molecular ion interference was determined. Spike concentrations ranged from 5 to 30 µg/L for trace elements and 0.5 mg/L for Ca, K, Mg, Na, and Si. Concentrations ranged from 2 mg/L for interfering trace elements like Mo to 500 mg/L for major elements like Ca. Based on the percentage spike recovery results, only Co, Fe, Ni, and Zn required the use of a cell gas to eliminate molecular ion interference (see table 4). For example, if a cell gas was not used, 3 µg/L of Co would be measured in the presence of 500 mg/L Ca because of the $^{43}Ca^{16}O^+$ interference. Molecular ions $^{29}Si^{27}Al^+$ and $^{40}Ca^{16}O^+$ severely affected the determination of Fe when hydrogen was not used making it difficult to estimate the degree of interference. All the other interference solutions indicated negligible bias for the interferent concentration range tested. For some elements where negligible bias was indicated, no cell gas is used during routine analyses (table 2) because the interferent concentration tested is rarely exceeded (for example, the interferences on Ag and Cd, see table 3), or the element being measured is normally present at concentrations higher than trace levels (for example, Al and Mn). Nevertheless, the use of a cell gas might be required when interferent concentrations are greater than those tested.

Interferences also may be associated with aqueous sample introduction and ionization processes. Dissolved-solid concentrations affect nebulization efficiency and suppress ionization. Typical limitations for dissolved-solid concentration for most pneumatic nebulizers range from 0.1 to 0.3 percent; some high dissolved-solid nebulizers do not have this limitation but also have higher sample flow rates. High dissolved-solid concentrations associated with easily ionized cations like sodium affect the ionization of analyte ions. Dissolved-solid concentrations of 0.1 to 0.3 percent translate into estimated specific conductances of between 4,000 and 6,000 µS/cm in natural water (Hem, 1986). Percentage recoveries in sample matrices with increasing specific conductance were used to establish the effect of specific conductance on cICP–MS analyses for various elements extending the mass range (see figure 4). The variability for elements heavier than ^{51}V averaged less than 3 percent over the duration of the experiment (about 17 hours); the variability for lighter elements ranged from 5 to 8 percent. These test-sample results suggest that samples having specific conductances less than 7,500 µS/cm can be analyzed directly with acceptable bias and variability. In previous ICP–MS methods, samples with specific conductances greater than 2,500 µS/cm required dilution (Faires, 1993; Garbarino and Taylor, 1994; Garbarino and Struzeski, 1998; Garbarino, 1999, 2000). The median specific conductance for samples submitted to NWQL in 2004 was

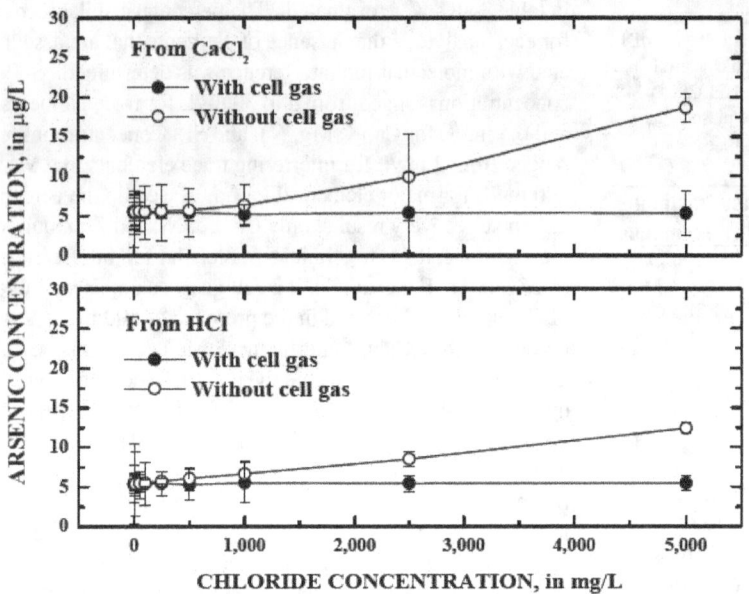

Figure 1. The effect of increasing concentrations of chloride from two sources on the determination of arsenic at mass-to-charge ratio 75. The graphs show that by using helium as the cell gas the molecular ion interference from chloride [$^{40}Ar^{35}Cl^+$] is minimized. The chloride concentration was about equal in each source. Error bars correspond to the standard deviation for three replicate determinations. Arsenic and chloride concentrations are in micrograms per liter (µg/L) and milligrams per liter (mg/L), respectively.

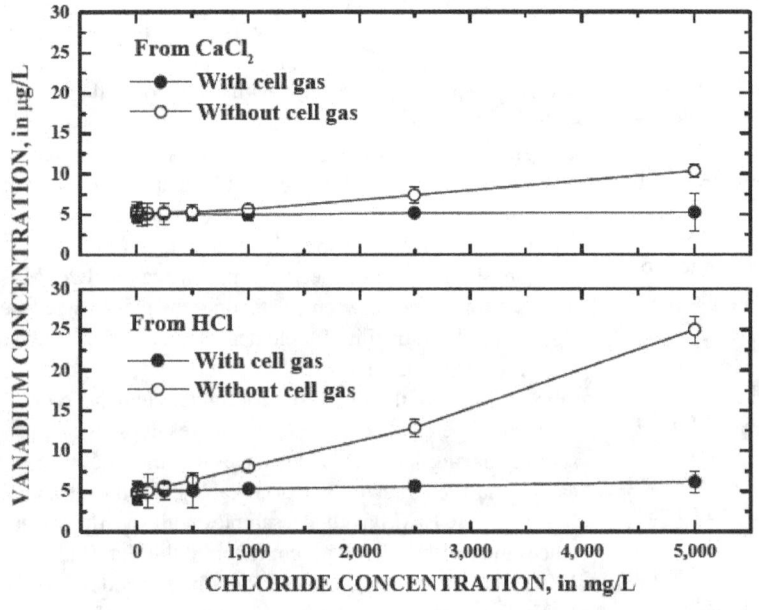

Figure 2. The effect of increasing concentrations of chloride from two sources on the determination of vanadium at mass-to-charge ratio 51. The graphs show that by using helium as the cell gas the molecular ion interference from chloride [$^{35}Cl^{16}O^+$] is minimized. The chloride concentration was about equal in each source. Error bars correspond to the standard deviation for three replicate determinations. Vanadium and chloride concentrations are in micrograms per liter (µg/L) and milligrams per liter (mg/L), respectively.

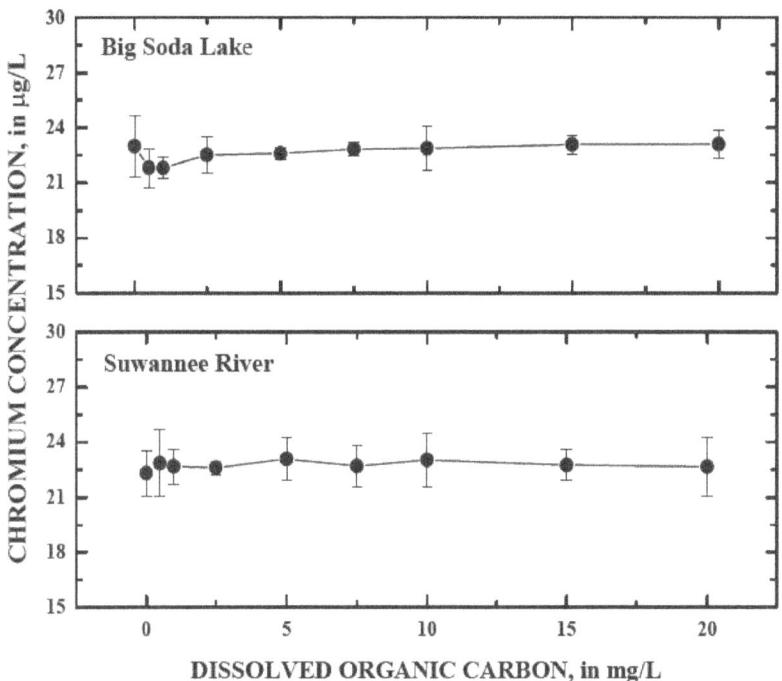

Figure 3. The effect of increasing concentrations of dissolved organic carbon from two natural-water sources on the determination of chromium at mass-to-charge ratio 52. The graphs show that by using helium as the cell gas the molecular ion interference from carbon [$^{40}Ar^{12}C^+$] is minimized. Error bars correspond to the standard deviation for three replicate determinations. Chromium and dissolved organic carbon concentrations are in micrograms per liter (µg/L) and milligrams per liter (mg/L), respectively.

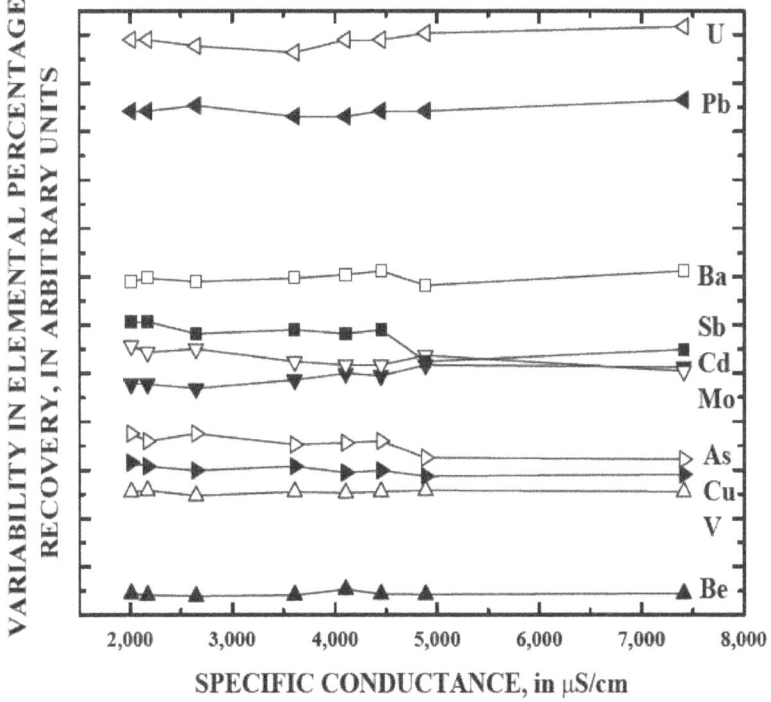

Figure 4. The variability in percentage recovery for selected elements extending the operational mass range as a function of increasing specific conductance. Specific conductances are in microsiemens per centimeter (µS/cm) at 25 degrees Celsius. The variability for elements heavier than vanadium averaged less than 3 percent over the range of specific conductance; the variability for lighter elements ranged from 5 to 8 percent. The data points represent the average for spiked natural-water samples analyzed in triplicate over a 17-hour period; the average standard deviation was 2.6 percent for elements with masses greater than vanadium and 16 percent for lighter elements.

Table 3. Potential molecular ion interferences for elements determined using collision/reaction cell inductively coupled plasma–mass spectrometry.

[All superscripted numbers are nominal mass-to-charge ratios]

Ion	Molecular ion	Ion	Molecular ion	Ion	Molecular ion
$^{23}Na^+$	$^7Li^{16}O^+$	$^{52}Cr^+$	$^{40}Ar^{12}C^+$ $^{35}Cl^{16}OH^+$ $^{36}Ar^{16}O^+$ $^{51}VH^+$	$^{75}As^+$	$^{40}Ar^{35}Cl^+$ $^{48}Ti^{27}Al^+$ $^{63}Cu^{12}C^+$
$^{24}Mg^+$	$^{12}C_2^+$ $^{23}NaH^+$	$^{55}Mn^+$	$^{40}Ar^{14}NH^+$ $^{28}Si^{27}Al^+$ $^{39}K^{16}O^+$ $^{54}FeH^+$ $^{54}CrH^+$	$^{78}Se^+$	$^{77}SeH^+$ $^{40}Ar^{38}Ar^+$ $^{66}Zn^{12}C^+$
$^{27}Al^+$	$^{12}C^{14}NH^+$ $^{26}MgH^+$ $^{11}B^{16}O^+$ $^{13}C^{14}N^+$ $^{12}C^{15}N^+$	$^{56}Fe^+$	$^{40}Ar^{16}O^+$ $^{55}MnH^+$ $^{40}Ca^{16}O^+$ $^{29}Si^{27}Al^+$ $^{44}Ca^{12}C^+$ $^{40}K^{16}O^+$	$^{107}Ag^+$	$^{91}Zr^{16}O^+$
$^{28}Si^+$	$^{14}N_2^+$ $^{27}AlH^+$ $^{12}C^{16}O^+$	$^{59}Co^+$	$^{58}FeH^+$ $^{43}Ca^{16}O^+$ $^{41}K^{18}O^+$	$^{111}Cd^+$	$^{95}Mo^{16}O^+$
$^{39}K^+$	$^{38}ArH^+$ $^{27}Al^{12}C^+$ $^{23}Na^{16}O^+$	$^{60}Ni^+$	$^{44}Ca^{16}O^+$ $^{48}Ca^{12}C^+$ $^{23}Na^{37}Cl^+$ $^{43}Ca^{16}OH^+$	$^{182}W^+$	$^{166}Er^{16}O^+$
$^{40}Ca^+$	$^{40}Ar^+$ $^{28}Si^{12}C^+$ $^{24}Mg^{16}O^+$ $^{40}K^+$	$^{63}Cu^+$	$^{40}Ar^{23}Na^+$ $^{31}P^{16}O_2^+$ $^{51}V^{12}C^+$ $^{47}V^{16}O^+$		
$^{51}V^+$	$^{35}Cl^{16}O^+$ $^{39}K^{12}C^+$	$^{66}Zn^+$	$^{34}S^{16}O_2^+$ $^{33}S_2^+$ $^{54}Fe^{12}C^+$ $^{50}Cr^{16}O^+$ $^{50}Ti^{16}O^+$ $^{54}Cr^{12}C^+$		

Table 4. Effectiveness of using a collision/reaction cell gas for the elimination of molecular ion interferences.

[Analyte, nominal elemental concentration; Interferent(s), elements associated with molecular ions that interfere with the analyte and the nominal concentration tested; Cell gas, N identifies an analyte that does not require the use of a cell gas for the interferent concentration tested and X identifies an analyte that requires a cell gas for accurate quantitation followed in parentheses by either the apparent analyte concentration, in micrograms per liter, associated with interferent concentration listed or an *s* indicating that the apparent analyte concentration could not be estimated because of severe molecular ion interference; mg/L, milligrams per liter; &, and]

Analyte	Interferent(s)	Cell gas	Analyte	Interferent(s)	Cell gas
Ag	Zr, 2 mg/L	N	Mg	Na, 500 mg/L	N
				C, 50 mg/L	N
Al	Mg, 200 mg/L	N			
	C & N, 50 mg/L	N	Mn	K, 200 mg/L	N
				Fe, 500 mg/L	N
Ca	Mg, 200 mg/L	N			
	K, 200 mg/L	N	Ni	Ca, 500 mg/L	X(14)
	C & Si, 50 mg/L	N		Ca & C, 100 mg/L	X(13)
				Na, 500 mg/L	N
Cd	Mo, 2 mg/L	N		Na & Cl, 500 mg/L	N
Co	Ca, 500 mg/L	X(3)	Se	C & Zn, 50 mg/L	N
	Fe, 500 mg/L	N			
	K, 200 mg/L	X(*s*)	Si	Al, 200 mg/L	N
				C, 50 mg/L	N
Cu	Na, 500 mg/L	N			
	P, 200 mg/L	N	V	K & C, 100 mg/L	N
				C & K, 50 mg/L	N
Cr	C, 50 mg/L	N			
			W	Er, 2 mg/L	N
Fe	Al & Si, 200 mg/L	X(*s*)			
	Ca, 500 mg/L	X(*s*)	Zn	Fe & C, 100 mg/L	N
	Ca & C, 100 mg/L	X(8)		S, 200 mg/L	N
	K, 200 mg/L	X(*s*)		C & Fe, 50 mg/L	N
	C & Ca, 50 mg/L	N		Ti, 2 mg/L	X(2)
K	Al & C, 100 mg/L	N			
	Na, 500 mg/L	N			

398 µS/cm; 95 percent of the samples had specific conductances of less than 3,000 µS/cm.

Instrumentation

The instrumental operating conditions for the cICP–MS method are listed in table 5. The inductively coupled plasma operating characteristics are typical of those used in earlier methods. However, there are substantial differences in the sample introduction and ion optics in conjunction with the use of the collision/reaction cell. A pneumatic concentric nebulizer is used to introduce samples at a flow rate of 0.2 to 0.3 mL/min into a thermostatically controlled spray chamber. This introduction system minimizes sample volume requirements, increases sample introduction efficiency, reduces instrumental drift, and reduces oxide and hydroxide molecular ions. The internal standards are introduced automatically through a junction tee (see fig. 5A). Most of the lens potentials and pole biases are similar regardless of whether a collision gas is introduced; notable differences are the cell entrance and exit potentials and the quadrupole (QP) and octapole (OctP) biases (see table 5). Nominal hydrogen and helium flow rate in the gas cell is about 4 mL/min. Previous ICP–MS methods used substantially different sample-introduction systems (nebulizer and spray chamber) and ion optics (see Faires, 1993; Garbarino and Taylor, 1994; Garbarino and Struzeski, 1998; Garbarino, 1999 and 2000; Garbarino and others, 2002).

Minor changes were required for determining the distribution of arsenic species using high-performance liquid chromatography (HPLC) in combination with cICP–MS instrumentation. An Agilent 1100 HPLC was used to separate the arsenic species with various anion exchange columns and mobile phases as described in the original methods (Garbarino and others, 2002). No changes were needed for arsenic speciation methods that use hydride generation for sample introduction. However, there was a mismatch in flow rates between the HPLC (about 1 mL/min) and the microflow nebulizer (about 0.3 mL/min) for other arsenic speciation methods. The mismatch required using the junction tee that is usually used for introducing the internal standard to reduce flow into the nebulizer (see fig. 5B). Differences in data-acquisition characteristics, nebulizer flow rate, spray-chamber volume, and transfer-line lengths had negligible effects on the chromatographic resolution and retention times.

Refer to Agilent Technologies (2001, 2004a, 2004b) and NWQL Standard Operating Procedures INCM0458.0 (M.E. Cree, U.S. Geological Survey, written commun., 2005) and ID0359.1 (J.R. Garbarino, U.S. Geological Survey, written commun., 2005) for details of the operation and maintenance of the instrumentation used in this report.

Sample Collection, Preservation, Shipment, and Holding Times

Samples must be collected and preserved using the protocols outlined in the USGS National Field Manual for the Collection of Water-Quality Data (U.S. Geological Survey, variously dated). Samples used for the determination of dissolved elements in natural-water samples need to be filtered using either a 0.45-µm membrane capsule filter or in-line filter, and the filtrate must be preserved using contaminant-free nitric acid. Samples used for the determination of recoverable elements in unfiltered water also must be preserved with contaminant-free nitric acid and digested using the in-bottle procedure (Hoffman and others, 1996). The prescribed holding times for acid-preserved filtered and unfiltered samples stored in tightly capped polyethylene bottles for elemental determinations in this method is 6 months (U.S. Environmental Protection Agency, 2005).

Biota samples, including mollusks, fish, aquatic insects, crayfish, and aquatic plants, need to be collected and processed as outlined by Crawford and Luoma (1993). Samples are stored frozen in glass or polyethylene containers at –20°C. Frozen samples need to be shipped overnight on dry ice to the laboratory. The holding time for frozen samples analyzed for the elements determined in this method is 6 months from the time of collection (Crawford and Luoma, 1993). Biota samples are digested with nitric acid using a closed-vessel microwave digestion procedure (U.S. Environmental Protection Agency, 1996). The holding time for biota digestates has not been established but is expected to be the same as acid-preserved aqueous samples or sediment and soil digestates (6 months).

Sediment and soil samples are collected using coring or other suitable techniques, stored in glass or polyethylene containers, and shipped overnight on ice to the laboratory (U.S. Geological Survey, variously dated). In the laboratory, the samples are air-dried and sieved using a 2-mm plastic sieve (Fishman and Friedman, 1989). Sieved samples are stored in glass or polyethylene containers. The holding times for unprocessed and processed sediment and soil samples have not been established because total elemental concentrations will most likely not change. Various subsampling techniques, such as cone-and-quartering or riffle splitting, are used to obtain a representative subsample of the sieved material prior to digestion. Samples are digested with nitric acid using a closed-vessel microwave digestion procedure (U.S. Environmental Protection Agency, 1998); the digestion is typically not a total dissolution, therefore the results represent recoverable concentrations. The recommended holding time for sediment and soil digestates is 6 months (U.S. Environmental Protection Agency, 2005).

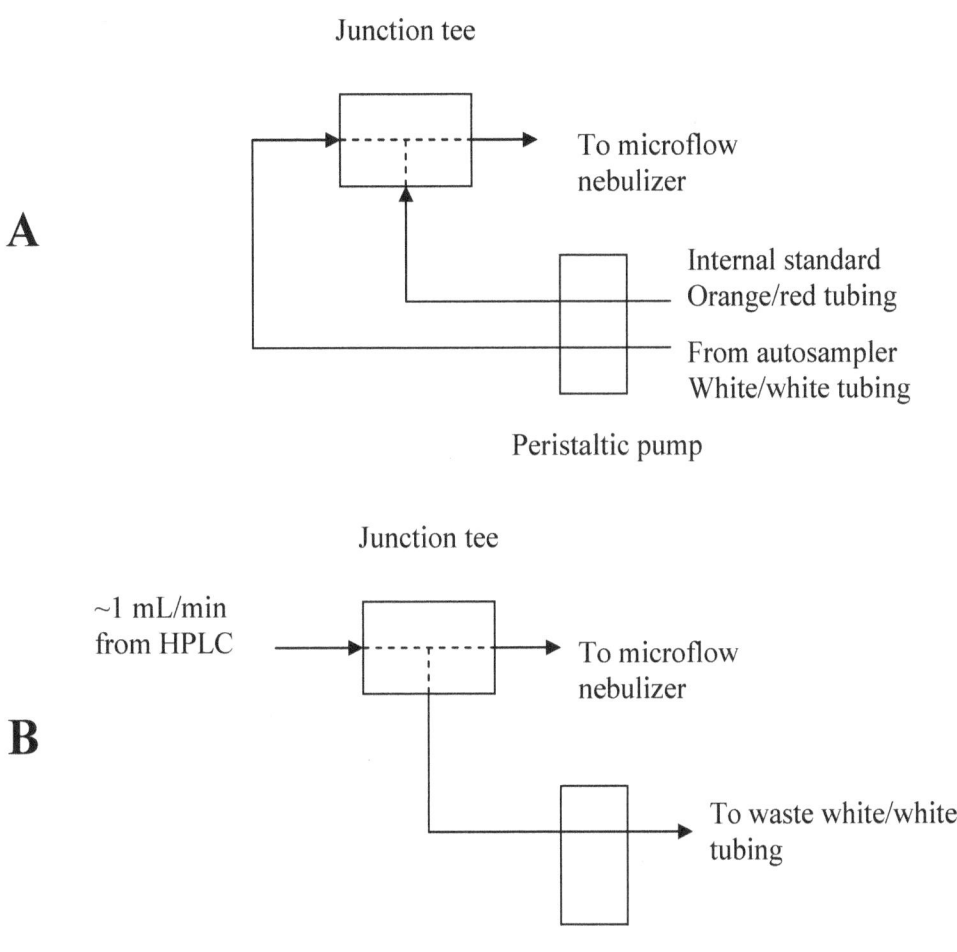

Junction tee

To microflow
nebulizer

Internal standard
Orange/red tubing

From autosampler
White/white tubing

A

Peristaltic pump

Junction tee

~1 mL/min
from HPLC

To microflow
nebulizer

To waste white/white
tubing

B

Peristaltic pump

Figure 5. Sample-flow paths used with new collision/reaction cell inductively coupled plasma–mass spectrometric methods. Panel A represents the sample-flow path used for the introduction of internal standard solution with filtered samples and filtered biota, sediment, and soil digestates. Panel B represents the sample-flow path used for arsenic speciation that reduces the flow rate from the high-performance liquid chromatograph (HPLC) to the level that is required by the microflow nebulizer. The pump speed is increased over that used in panel A so that the flow rate at the nebulizer is reduced from 1 milliliter per minute (mL/min) to about 0.3 mL/min.

Table 5. Typical operating characteristics for collision/reaction cell inductively coupled plasma–mass spectrometry.

[cICP–MS, collision/reaction cell inductively coupled plasma–mass spectrometer; Pa, pascal; °C, degrees Celsius; %, percent; L/min, liters per minute; kPa, kilopascal; W, watts; mL/min, milliliters per minute; rpm, revolutions per minute; mm, millimeters; id, inside diameter; V, volts; <, less than; OctP, octapole; QP, quadrupole; RF, radio frequency]

cICP–MS	Typical setting (range)		
Instrument	Agilent 7500ce		
Forward power	1550 W (700–1600)		
Reflected power	4 W (<20)		
Analyzer pressure	3.33×10^{-3} Pa (3×10^{-4} to 2×10^{-3})		
Carrier argon	1.00 L/min (0.8 to 1.3)		
Makeup argon	None (0 to 1.0 L/min)		
Plasma argon	15 L/min (about 15)		
Auxiliary argon	0.9 L/min (0 to 1.0)		
Sampling depth	8 mm (7 to 10)		
Sample introduction			
Concentric nebulizer	0.3 mL/min (0.2 to 0.3)		
Quartz spray chamber	2°C		
Peristaltic pump	0.1 rpm, during data acquisition		
Sample tube	white/white, 1.02 mm id		
Internal standard tube	orange/red, 0.19 mm id		
Waste tube	yellow/blue, 1.52 mm id		
Ion optics and collision/reaction cell			
Extract 1	0 V		
Extract 2	–133 V (–120 to –150)		
Omega bias	–16 V (–15 to –40)		
OctP RF	161 V (160 to 200)		
	Normal mode	He mode	H_2 mode
Cell gas	0 mL/min	4.0 (2.5 to 5.5)	4.0 (3 to 6)
Cell entrance	–30 V (–30 to –10)	–25	–25
QP focus	3 V (1 to 5)	–8	–8
Cell exit	–30 V (–30 to –8)	–45	–45
OctP bias	–6 V (–10 to –3)	–18	–18
QP bias	–3 V (–7 to 0.5)	–13	–5

Filtered arsenic speciation samples are collected using the procedures outlined in the USGS National Field Manual for the Collection of Water-Quality Data (U.S. Geological Survey, variously dated). Samples are preserved with ethylenediaminetetraacetic acid and stored in opaque polyethylene bottles (U.S. Geological Survey, variously dated, specifically chap. A5.6.4). Samples that are submitted for speciation in the laboratory need to be shipped on ice and analyzed within 90 days of collection to ensure the accuracy of the results. The solid phase extraction cartridge and eluate for samples that are speciated in the field need to be shipped on ice to the laboratory within 30 days of collection. The cartridge needs to be extracted in the laboratory within 49 days of sample collection. The extract and eluate need to be analyzed within 6 months.

Calibration Standards and Other Solutions

ASTM Type I reagent water (American Society for Testing and Materials, 2000, p. 10), acids, and reagents used in this method must be verified to have analyte concentrations less than the method detection limits. Calibration standards for elemental determinations are prepared from either commercially available single-element stock solutions or multi-element stock solutions. The concentrations in all solutions used need to be verified by using National Institute of Standards and Technology (NIST) standards or standards that are traceable to NIST as a reference. Multielement calibration standards containing Ag, Al, As, B, Ba, Be, Ca, Cd, Co, Cr, Cu, Fe, K, Li, Mg, Mn, Mo, Na, Ni, Pb, Sb, Se, Si, Sr, Tl, U, V, W, and Zn are prepared by making appropriate dilutions of commercial stock solutions with reagent water acidified with nitric acid (0.4 percent by volume). Calibration standards for whole-water digestates are prepared in 0.4 percent nitric acid and 2 percent hydrochloric acid. The multielement standard containing Sb also must contain 0.1 percent hydrofluoric acid by volume as a preservative. Calibration concentrations are typically 10 and 100 µg/L for all trace elements, whereas Al, Ba, Mn, Sr, and Zn use an additional standard at 2,500 µg/L to extend the calibration range. Multielement calibration standards for Ca and Fe were prepared at 25 and 250 mg/L and 100 and 1,000 µg/L, respectively, whereas K, Mg, Na, and SiO_2 were prepared at 10 and 100 mg/L.

Several internal standard elements are introduced automatically into the sample stream to minimize sample matrix effects and instrumental drift (see fig. 5A). A multielement internal standard solution containing Li (highly enriched in lithium-6), Bi, Ce, Cs, Ge, In, Rh, Sc, Th, and Y at 1 mg/L is prepared by making appropriate dilutions of each stock solution with reagent water acidified with nitric acid (0.4 percent by volume).

The cICP–MS operating conditions are optimized daily by using a tuning solution composed of Ce, Co, Li, Mg, Th,

and Y at 10 µg/L each. The tuning solution was prepared by making appropriate dilutions of each stock solution with reagent water and acidifying with nitric and hydrochloric acid at 0.4 and 2 percent (by volume), respectively.

The electron multiplier of the cICP–MS is operated in the pulse-counting mode whenever an elemental concentration produces less than about 2 million counts per second. Higher count rates saturate the multiplier, thereby requiring that the multiplier be used in the analog mode. The linear transition between the two modes requires that a pulse/analog (P/A) calibration factor be established daily for each element determined. A solution containing all the elements is prepared by making the appropriate dilution of the stock solutions with reagent water acidified with nitric acid (0.4 percent by volume). Elemental concentrations range from 5 to 25 µg/L.

The performance of cICP–MS in interference reduction is tested periodically by analyzing solutions containing major interfering constituents and affected analytes. A solution having interfering constituents contains 1,000 mg/L Cl (from HCl); 100 mg/L each of Al, K, Mg, Mn, P, S, and Zn; 50 mg/L Si; 250 mg/L each of C (from potassium hydrogen phthalate), Ca, Fe, and Na; and 2 mg/L each of Mo, Ti, and Er. Another solution that also contains selected analytes is prepared as above with the addition of Ag, As, Cd, Co, Cr, Cu, Ni, Se, V, and W, each at 20 µg/L. Potential interferences associated with these elements are listed in table 3. The solutions are prepared by making suitable dilutions of contaminant-free commercial solutions (except for C and Cl) with reagent water acidified to 1 percent by volume with nitric acid.

Quantitation

Analyte concentrations are calculated automatically from the linear regression equations by the operating system. The operating system monitors results for quality-control and quality-assurance samples in real-time and updates calibration whenever acceptance criteria fail. Furthermore, sample carryover is minimized by automatically increasing the length of the rinse-cycle period until an analyte signal reaches an acceptable level. Such automatic quantitation and control features are not unlike those provided by older ICP–MS instruments in methods for routine elemental analyses. However, He or H_2 must be introduced into the cell during data acquisition for elements that are affected by molecular ion interferences and purged from the cell for other elements. There is also a notable difference in the procedures used to calculate arsenic speciation results when using the cICP–MS. Previous arsenic speciation methods were unable to calculate results in real time. The chromatographic peaks were integrated and the regression equations were calculated offline and required substantial manual data processing. The new operating system software makes it possible to process the results and calculate concentrations in real time.

Reporting of Analytical Data

Elemental concentrations presented in USGS data reports are reported so that the rightmost digit (called the least significant digit) represents the uncertainty in the result (Novak, 1985; Hansen, 1991; U.S. Geological Survey, 2002). The least significant digit is determined using guidance outlined by the American Society for Testing and Materials (1999). NWQL reports results in the national database to the least significant digit plus one additional digit. Results for water samples will be reported in either micrograms per liter or milligrams per liter, depending on the element (see table 1). Elemental results for biota, sediment, and soil digestates will be reported in micrograms per gram or milligrams per gram, dry weight, depending on the element. All arsenic species are reported in micrograms–arsenic per liter.

Laboratory Quality Assurance/Quality Control

Laboratory samples are prepared in sample sets consisting of the calibration standards, the interference check solution, laboratory reagent blanks (LRB), and laboratory control samples (LCS); the required frequency of analyzing LRB and LCS samples is outlined below. Furthermore, standard reference materials, laboratory reagent spike samples, and laboratory duplicate samples can be included to meet specific data-quality objectives. The NWQL uses the Quality Management System Document for essential Quality-Assurance/Quality-Control requirements (Maloney, 2005).

The LRB is prepared to match the matrix of the samples and calibration standards. A minimum of one LRB is analyzed for every 10 analytical samples. The LRB cannot have any detectable levels of contaminants. The control limits for the LRB are plus or minus the LT–MDL for each element. For LRB that have a detectable concentration of a certain element, the source of contamination must be determined and either eliminated or minimized before the environmental samples are reanalyzed. In cases where the source of contamination cannot be identified, the element is assigned a data-qualifying code. The concentration of the element in the environmental sample, however, can be reported without a qualifying code if it is 10 times greater than the concentration in the LRB. Conversely, if concentrations are less than the LT–MDL, the samples need to be reanalyzed.

The LCS is prepared from either a solution having a different lot number than the calibration standards or a solution from a different vendor than the calibration standards. In either case, the solution used to prepare the LCS needs to be traceable to NIST. A minimum of one LCS is analyzed for every 10 analytical samples. The control limits for the LCS are plus or minus two standard deviations of the expected concentration of the element of interest. Any elemental results that fall outside of the control limits are not reported. All environmental samples associated with the failed LCS must be reanalyzed until the specified acceptance criterion is met.

All material used to make calibration standards need to be NIST traceable. Calibration standards are verified by using a NIST traceable LCS. Determined concentrations need to be within two standard deviations of the specified concentration.

Results and Discussion of Method Validation Data

The analytical bias and variability across the normal calibration range of the analytes were determined by using results for two USGS standard reference water samples (SRWS), and results for reagent-water, ground-water, and surface-water samples spiked at several concentrations. The SRWSs are based on spiked natural-water matrices whose most probable elemental concentrations have been established through interlaboratory analyses using multiple analytical methods. SRWS T145 was made with water collected from the Yampa River near Steamboat Springs, Colorado (Farrar and Long, 1997), and SRWS T169 was made with water collected from the South Platte River west of Bailey, Colorado (Woodworth and Conner, 2002); water samples from both sources were subsequently filtered, spiked, and preserved with nitric acid. SRWS T145 and T169 were analyzed between 21 and 68 times over several days and multiple instrument calibrations. A statistical summary of these results is listed in table 6. The average results for T145 are within one F-pseudosigma of the most probable concentration except for Na and Mg. Nevertheless, the averages for Na and Mg are within 1.5 F-pseudosigma of the most probable concentration, which is considered to be acceptable. The average results for T169 are within one F-pseudosigma of the most probable concentration except for Ba, Ni, and U, which are within 1.5 F-pseudosigma. The concentration of W in the SRWSs was not established; therefore, a NIST W standard was used to prepare solutions in reagent water. The resulting experimental W concentrations were within 4 and 18 percent of the expected concentrations of 1 and 50 μg/L, respectively. The corresponding variability ranged from 3 to 7 percent for 30 measurements. Results for reagent blank samples analyzed with the SRWSs showed acceptably low concentration levels with slight variability (see table 6).

The new cICP–MS method was validated by determining the percentage bias and variability at 2 to 4 spike-concentration levels in reagent-water, ground-water, and surface-water matrices. Reagent water was spiked at 4 concentrations; 2 levels were in the lower one-third of the normal calibration range, 1 at roughly one-third and 1 at roughly two-thirds of the normal calibration range. Ground-water and surface-water matrices were spiked at roughly one-third and two-thirds of the normal calibration range. The number of measurements at each concentration was between 16 and 33 for each matrix determined over several days.

Table 6. Long-term bias and variability of results from aqueous standard reference materials using collision/reaction cell inductively coupled plasma–mass spectrometry.

[Analyte, concentrations are in micrograms per liter unless otherwise noted; SRWS, U.S. Geological Survey standard reference water sample; Exp, mean concentration plus or minus one standard deviation for experimental results based on the number of measurements, in italics, measured over several days for SRWS T145 and T169; MPV, most probable mean value plus or minus one F-pseudosigma for interlaboratory results using multiple methods except for tungsten (W), which are nominal concentrations from a diluted National Institute of Standards and Technology reference standard; Blank, mean concentration plus or minus one standard deviation for 35 reagent blank measurements taken over several days; mg/L, milligrams per liter]

Analyte	SRWS T145		SRWS T169		Blank
	Exp	MPV	Exp	MPV	
Ag	7.8±0.3, *68*	7.6±0.9	3.9±0.1, *49*	3.9±0.2	−0.01±0.01
Al	64±6, *52*	68±11	32.9±0.6, *21*	34±4	−0.3±0.2
As	10.4±0.6, *52*	10±1	8.6±0.7, *49*	8.6±0.7	−0.02±0.03
B	46±2, *40*	45±6	25±1, *49*	24±2	0.3±0.6
Ba	36±2, *68*	37±2	41±1, *49*	43±2	0.05±0.1
Be	9.0±0.3, *40*	9.0±0.7	9.4±0.4, *49*	9.6±0.6	−0.009±0.01
Ca, mg/L	30±2, *28*	31±1	37±1, *21*	37.9±0.9	−0.02±0.02
Cd	9.8±0.3, *68*	9.3±0.8	3.60±0.08, *49*	3.4±0.2	−0.007±0.01
Co	10.0±0.6, *68*	10.0±0.9	1.91±0.05, *49*	1.9±0.2	−0.002±0.04
Cr	13.9±0.5, *48*	15±1	7.8±0.3, *49*	8.2±0.4	−0.024±0.02
Cu	11.6±0.6, *68*	11±1	14±1, *49*	14.3±0.8	0.008±0.1
Fe	94±4, *38*	101±8	12.1±0.5, *21*	11±4	−0.03±0.2
K, mg/L	2.0±0.1, *28*	2.1±0.2	2.50±0.07, *21*	2.6±0.1	−0.014±0.02
Li	26.6±0.9, *40*	27±2	9.1±0.3, *49*	9.6±0.6	0.003±0.04
Mg, mg/L	8.1±0.5, *28*	8.7±0.4	4.2±0.1, *21*	4.3±0.1	−0.02±0.03
Mn	20±1, *68*	21±2	27±1, *49*	27.0±0.9	0.02±0.08
Mo	8.6±0.4, *68*	9±1	71.8±0.6, *25*	71±2	−0.005±0.02
Na, mg/L	39±2, *21*	41±2	10.3±0.4, *21*	10.6±0.3	−0.02±0.02
Ni	11±1, *68*	11±1	11.1±0.6, *49*	10.3±0.7	−0.11±0.03
Pb	12.4±0.6, *68*	13±1	10.7±0.3, *49*	11.4±0.7	−0.01±0.02
Sb	8.3±0.4, *68*	8.8±0.9	2.92±0.09, *49*	3.3±0.4	−0.008±0.01
SiO_2, mg/L	11.5±0.8, *28*	11.3±0.7	6.2±0.2, *21*	6.0±0.2	−0.03±0.05
Se	10.7±0.8, *52*	10±1	2.7±0.2, *49*	2.9±0.4	−0.07±0.08
Sr	208±7, *68*	203±19	176±6, *49*	174±6	0.2±0.2
Tl	15.1±0.8, *68*	15±3	4.3±0.1, *49*	4.8±0.5	−0.04±0.02
U	1.02±0.06, *68*	1.1±0.1	1.55±0.02, *25*	1.7±0.1	−0.008±0.009
V	11.2±0.7, *68*	12±2	7.4±0.3, *49*	7.4±0.4	−0.03±0.02
W	0.96±0.07, *30*	1	59±2, *30*	50	−0.1±0.3
Zn	9.6±0.8, *68*	10±2	18±1, *49*	19±1	−0.2±0.4

The bias and variability for reagent water spiked with analyte concentrations in the lower one-third of the normal calibration range are listed in table 7. The percentage bias for reagent water spiked at trace-element concentrations of 0.5 to 3 μg/L (column 3 of table 7) averaged 4.4 percent with a range of –6 to 16 percent, whereas the average percentage bias for Ca, K, Mg, Na, and SiO_2 was 1.4 percent with a range of –4 to 10 percent for spikes of 0.5 to 3 mg/L. For the other 3 spike levels (column 6 of table 7 and columns 3 and 6 of table 8), the average bias for trace elements was 0.9 percent with percentages ranging from –8.1 to 13 percent, whereas for Ca, K, Mg, Na, and SiO_2, the average bias was 1.0 percent with percentages ranging from –7.7 to 10 percent. The corresponding variability ranged from 0.7 to 10 percent relative standard deviation (percent RSD) for trace elements and from 1.0 to 10 percent RSD for Ca, K, Mg, Na, and SiO_2.

Validation data for surface-water and ground-water matrices are listed in tables 9 and 10, respectively. Surface water was spiked with between 17 and 1,754 μg/L for trace elements and 8 and 95 mg/L for Ca, K, Mg, Na, and SiO_2. The average bias for trace elements was 1.8 percent with percentages ranging from –14 to 21 percent; for Ca, K, Mg, Na, and SiO_2, the average bias was 0.96 percent with a range of –4.2 to 7.8 percent. The associated variability ranged from 2.6 to 22 percent RSD for trace elements and from 1.0 to 5.8 percent RSD for Ca, K, Mg, Na, and SiO_2. Ground water was spiked with between 17 and 1,769 μg/L for trace elements and 9 and 95 mg/L for Ca, K, Mg, Na, and SiO_2. The average bias for trace elements was –0.59 percent with percentages ranging from –16 to 16 percent; for Ca, K, Mg, Na, and SiO_2, the average bias was 1.5 percent with a range of –7.5 to 11 percent. The associated variability ranged from 1 to 18 percent RSD for trace elements and from 0.8 to 5.0 percent RSD for Ca, K, Mg, Na, and SiO_2.

Linear regression analysis was used to evaluate whether cICP–MS results for 25 filtered natural-water samples and 25 unfiltered natural-water digestates were notably different than results from established analytical methods. The elemental concentration range for the test samples typically extended 3 to 4 orders of magnitude (see column 6 in tables 11 and 12, in addition, see graphs of regression analyses in the Appendix). The slopes of the regression equations were used to identify systematic or random errors in the analytical results; a slope of unity indicates there is no difference between instrumental methods. A nonzero y-intercept indicates possible faulty background correction for one of the methods.

Eighty-three percent (24 of 29) of the elements had slopes within ±10 percent of unity in the 95 percent confidence interval for filtered natural-water samples. Four of the remaining five elements were within ±15 to 30 percent; cICP–MS results for Li tended to be lower than the previous method and Tl and W tended to be higher. The slope for B was about 30 percent greater than unity possibly from carryover from samples with high B concentrations. The slope for Be was notably different than unity because of the low concentrations present in the test samples. For unfiltered digestate samples, 79 percent (23 of 29) of the elements had slopes within ±10 percent of unity in the 95 percent confidence interval. Four of the remaining six elements were within ±15 to 20 percent; cICP–MS results for Al, Cu, and W tended to be lower than the other methods and Mo tended to be higher. Slopes for Ag and Sb were substantially different than unity because of the low concentrations present in the test samples.

The median y-intercept for trace elements corresponding to the linear regression analysis of filtered results was 0.01 μg/L with a range from –1.7 to 69 μg/L; the median for Ca, K, Mg, Na, and SiO_2 was 0.05 mg/L with a range of –0.8 to 0.3 mg/L. B, Ba, Li, and Sr had the largest y-intercepts ranging from –2 to 69 μg/L and corresponded to sample results extending from 100 to 1,000 μg/L. For trace elements in whole-water digestates, the median y-intercept was 0.04 μg/L with a range from –26 to 1.7 μg/L; the medium for Ca, K, Mg, Na, and SiO_2 was –0.2 mg/L with a range of –0.6 to 1.1 mg/L. The y-intercepts for Al and Fe were –22 and –26 μg/L, respectively, most likely an effect of sample concentrations extending to 10,000 μg/L.

The bias and variability associated with the analysis of biological standard reference materials also was determined. The analysis of sediment or soil standard reference material was not tested because the microwave digestion procedure used by NWQL does not completely solubilize such solids, thereby making the comparison of results to certified concentrations problematic. The results of the analysis of National Research Council Canada Dolt-2 and Dorm-2 standard reference materials (SRM) and NIST SRM 1577b using cICP–MS are listed in table 13. Linear regression analysis comparing the experimental elemental concentrations from cICP–MS to the certified concentrations in the standard reference materials is shown in figure 6. The regression coefficients indicate an acceptable linear relation. Nevertheless, on the average, the experimental results were about 17 percent higher for elemental concentrations ranging five orders of magnitude.

Table 7. Long-term bias and variability for the analysis of reagent water spiked with analyte concentrations in the lower one-third of the normal calibration range using collision/reaction cell inductively coupled plasma–mass spectrometry.

[Spiked reagent-water samples were analyzed randomly with spiked surface- and ground-water samples over several days; Analyte, concentrations are in micrograms per liter unless otherwise noted; Exp, mean concentration plus or minus standard deviation for experimental results based on 16 to 22 replicate determinations at each spike concentration; Percent bias, percentage error between the expected concentration and the mean experimental concentration; Percent RSD, percent relative standard deviation; mg/L, milligrams per liter]

Analyte	Exp	Percent bias	Percent RSD	Exp	Percent bias	Percent RSD
Ag	1.05±0.03	5.0	2.9	5.17±0.04	3.4	0.8
Al	2.91±0.09	16	3.1	13.9±0.6	11	4.2
As	1.04±0.05	4.0	4.8	5.21±0.09	4.2	1.7
B	2.4±0.6	2.8	25	12.1±0.9	–3.0	7.5
Ba	2.6±0.2	4.8	5.7	13.2±0.3	5.6	2.3
Be	0.50±0.01	0	2.0	2.51±0.04	0.4	1.6
Ca, mg/L	2.19±0.09	9.5	4.0	16.6±0.2	10	1.0
Cd	1.05±0.02	5.0	1.9	5.20±0.08	4.0	1.5
Co	1.06±0.03	6.0	2.8	5.26±0.04	5.2	0.8
Cr	1.00±0.02	0	2.2	4.7±0.3	–5.8	6.2
Cu	2.7±0.3	7.2	9.7	13.0±0.3	4.2	2.0
Fe	2.4±0.3	9.1	11	11.5±0.1	0	0.7
K, mg/L	0.54±0.01	8.0	2.2	2.20±0.08	10	3.5
Li	2.53±0.08	1.2	3.2	12.6±0.3	0.4	2.3
Mg, mg/L	0.48±0.01	–4.0	2.7	2.09±0.07	4.5	3.2
Mn	2.68±0.07	7.2	2.6	13.4±0.2	7.2	1.6
Mo	1.00±0.02	0	2.0	4.90±0.05	–2.0	1.0
Na, mg/L	0.48±0.01	–4.0	2.7	5.1±0.2	1.8	4.5
Ni	1.07±0.05	7.0	4.7	5.2±0.1	3.0	2.1
Pb	1.05±0.05	5.0	4.8	5.1±0.2	2.8	3.7
Sb	1.00±0.03	0	3.0	4.86±0.04	–2.9	0.8
Se	1.12±0.06	12	5.4	5.6±0.2	12	3.0
SiO_2, mg/L	2.9±0.2	–2.3	6.5	16.0±0.3	6.7	1.8
Sr	1.89±0.07	–5.5	3.7	13.0±0.7	3.8	5.2
Tl	0.51±0.02	2.0	3.9	2.65±0.07	6.0	2.6
U	1.05±0.04	5.0	3.8	5.2±0.2	4.6	4.0
V	1.01±0.04	1.0	4.0	5.1±0.1	1.2	2.0
W	0.79±0.02	–1.3	2.7	8.9±0.1	11	1.0
Zn	2.87±0.09	13	3.1	14.1±0.3	13	2.3

Table 8. Long-term bias and variability for the analysis of reagent water spiked with analyte concentrations at one-third and two-thirds of the normal calibration range using collision/reaction cell inductively coupled plasma–mass spectrometry.

[Spiked reagent-water samples were analyzed randomly with spiked surface- and ground-water samples over several days; Analyte, concentrations are in micrograms per liter unless otherwise noted; Exp, mean concentration plus or minus standard deviation for experimental results based on 20 to 33 replicate determinations at each spike concentration; Percent bias, percentage error between the expected concentration and the experimental mean concentration; Percent RSD, percent relative standard deviation; mg/L, milligrams per liter]

Analyte	Exp	Percent bias	Percent RSD	Exp	Percent bias	Percent RSD
Ag	34±2	0.6	4.8	72±4	7.7	6.1
Al	310±20	−7.3	5.6	640±20	−4.0	2.6
As	35±3	5.7	7.6	66±3	−1.5	4.8
B	82±8	−1.8	10	160±20	−4.3	10
Ba	86±3	3.0	3.0	173±4	3.7	2.3
Be	16.5±0.8	−1.2	4.8	35±1	5.1	3.6
Ca, mg/L	48.6±0.9	−2.9	1.8	83±2	−2.6	3.1
Cd	33.8±0.8	1.2	2.4	65±1	−2.1	2.2
Co	34±1	1.5	3.2	66±2	−0.6	2.5
Cr	182±2	5.8	1.2	340±30	2.2	8.8
Cu	85±7	2.2	8.7	170±10	1.2	7.1
Fe	371±3	3.1	0.7	717±7	−0.4	1.0
K, mg/L	17.6±0.2	−2.0	1.2	35±2	−2.0	5.2
Li	80±3	−4.2	3.9	170±5	1.9	2.7
Mg, mg/L	19.4±0.6	2.0	3.3	38±2	0.7	5.5
Mn	320±10	−3.2	3.1	640±20	−3.5	2.9
Mo	31±1	−6.3	3.5	62±2	−7.5	3.8
Na, mg/L	20±1	1.8	6.5	39±3	−3.0	7.8
Ni	33±2	−0.9	5.4	66±3	−0.9	4.2
Pb	34±2	2.3	4.4	65±2	−1.8	2.8
Sb	32±2	−3.4	7.7	61±2	−8.1	3.2
Se	35±1	5.6	4.0	69±2	3.2	2.6
SiO$_2$, mg/L	22±2	−1.7	10	37±3	−7.7	7.3
Sr	800±40	−4.1	4.3	1,620±80	−3.2	4.7
Tl	17.3±0.7	3.5	4.0	34±1	2.1	2.8
U	32±2	−2.8	5.5	63±3	−5.7	5.3
V	32±2	−5.0	7.2	63±1	−4.8	2.3
W	41±2	7.9	6.0	73±5	−4.0	6.3
Zn	340±20	2.9	5.0	660±30	−1.2	4.7

Table 9. Long-term bias and variability for the analysis of surface water spiked with analyte concentrations at one-third and two-thirds of the normal calibration range using collision/reaction cell inductively coupled plasma–mass spectrometry.

[Spiked surface-water samples were analyzed randomly with spiked reagent- and ground-water samples over several days; Analyte, concentrations are in micrograms per liter unless otherwise noted; Exp, mean concentration plus or minus standard deviation for experimental results based on 20 to 30 replicate determinations at each spike concentration; Percent bias, percentage error between the expected concentration and the experimental mean concentration; Percent RSD, percent relative standard deviation; mg/L, milligrams per liter

Analyte	Exp	Percent bias	Percent RSD	Exp	Percent bias	Percent RSD
Ag	34±2	3.0	6.5	57±2	−14	3.7
Al	350±60	4.7	18	770±170	16	22
As	33±2	−1.4	6.5	66±4	−0.8	6.4
B	130±10	21	7.8	230±20	19	9.9
Ba	122±6	8.8	4.6	201±8	3.1	4.0
Be	18±2	8.4	11	34±4	3.5	11
Ca, mg/L	59±2	7.5	4.0	91±5	−3.9	5.8
Cd	33±2	−1.5	6.0	66±4	0	5.9
Co	34±2	0.7	5.3	66±4	−0.3	5.6
Cr	160±10	−5.1	6.6	310±20	−6.9	7.9
Cu	93±8	8.5	8.9	170±20	−0.6	11
Fe	440±12	8.0	2.6	640±20	−3.4	3.4
K, mg/L	18.3±0.8	−3.8	4.3	36±1	7.8	3.7
Li	90±9	5.2	9.5	170±20	−0.9	9.8
Mg, mg/L	20.2±0.8	−4.0	4.2	38.8±0.7	2.1	1.8
Mn	320±20	−2.5	5.3	630±30	−6.0	4.1
Mo	34±2	1.7	7.2	73±5	9.5	7.5
Na, mg/L	14.5±0.7	3.6	5.0	32.3±0.3	−3.0	1.0
Ni	33±2	−0.1	6.3	67±4	0.1	5.9
Pb	33±2	−1.0	6.4	67±4	0.6	6.4
Sb	32±2	−3.0	5.1	63±4	−5.3	5.7
Se	36±2	9.2	5.3	66±4	−1.2	5.4
SiO$_2$, mg/L	8.6±0.2	7.5	2.8	32±1	−4.2	4.1
Sr	940±50	2.1	5.4	1,680±80	−4.4	4.9
Tl	16±1	−4.2	6.4	34±2	0.6	6.6
U	38±2	−9.4	4.7	68±3	2.2	4.4
V	33±2	0	5.9	69±4	3.6	5.8
W	43±1	14	3.3	76±3	1.2	3.5
Zn	320±20	−5.8	5.0	640±30	−6.1	4.9

Table 10. Long-term bias and variability for the analysis of ground water spiked with analyte concentrations at one-third and two-thirds of the normal calibration range using collision/reaction cell inductively coupled plasma–mass spectrometry.

[Spiked ground-water samples were analyzed randomly with spiked reagent- and surface-water samples over several days; Analyte, concentrations are in micrograms per liter unless otherwise noted; Exp, mean concentration plus or minus standard deviation for experimental results based on 20 to 32 replicate determinations at each spike concentration; Percent bias, percentage error between the expected concentration and the experimental mean concentration; Percent RSD, percent relative standard deviation; mg/L, milligrams per liter]

Analyte	Exp	Percent bias	Percent RSD	Exp	Percent bias	Percent RSD
Ag	33±2	−0.9	5.7	64±4	−4.1	6.3
Al	320±60	−5.7	18	700±100	9.1	18
As	32±2	−5.1	7.0	66±5	−1.5	7.0
B	106±8	16	7.8	200±20	14	7.6
Ba	134±6	5.3	4.4	214±9	1.6	4.4
Be	17±2	1.2	9.2	33±3	−1.5	8.8
Ca, mg/L	57±2	3.2	3.2	88±4	−7.5	5.0
Cd	32±2	−4.2	5.6	66±4	−1.2	5.5
Co	32±2	−3.1	5.3	65±4	−1.8	5.4
Cr	162±2	−3.3	1.0	300±20	−8.7	8.1
Cu	93±7	2.7	7.7	170±10	−4.0	8.5
Fe	420±40	7.7	8.8	550±30	−16	4.8
K, mg/L	18.9±0.8	−0.5	4.2	35±1	5.7	3.1
Li	85±7	2.0	8.2	160±10	−2.8	7.7
Mg, mg/L	15.3±0.4	2.0	2.4	37±1	11.0	3.5
Mn	310±20	−6.7	4.8	630±30	−5.2	5.1
Mo	34±2	1.2	6.8	74±5	10	7.2
Na, mg/L	17.5±0.6	2.9	3.3	32.6±0.3	−2.4	1.0
Ni	32±2	−3.1	7.3	66±4	−0.8	6.5
Pb	32±2	−4.2	6.0	66±4	−0.9	6.0
Sb	32±2	−3.1	5.1	63±4	−5.5	5.7
Se	34±1	3.2	4.3	63±3	−5.0	4.6
SiO$_2$, mg/L	9.40±0.08	4.3	0.8	32.0±0.5	−3.9	1.6
Sr	920±50	−1.7	5.3	1,800±100	1.9	5.8
Tl	16±1	−4.2	6.4	34±2	0.6	6.6
U	44±2	7.0	4.8	75±1	−0.1	1.6
V	33±2	0	5.9	69±4	3.5	5.8
W	39.9±0.8	5.0	2.0	64±2	−3.9	3.2
Zn	320±20	−5.8	5.0	640±30	−6.1	4.9

Table 11. Linear regression results for filtered natural-water samples analyzed using collision/reaction cell inductively coupled plasma–mass spectrometry and other spectrometric methods.

[Natural-water samples filtered using a 0.45-micrometer pore-size membrane and acidified to a pH of less than 2; Analyte, concentrations are in micrograms per liter unless otherwise noted; Slope, see Appendix for other spectrometric method used in comparison; 95% CI, 95-percent confidence interval for the slope or y-intercept; y-inter, y intercept; Conc. range, concentration range of the 25 samples tested; MDL, method detection limit (see table 2); mg/L, milligrams per liter]

Analyte	Slope	95% CI for slope	y-inter	95% CI for y-inter	Conc. range
Ag	0.933	0.929 to 0.936	−0.006	−0.022 to 0.009	MDL to 19
Al	0.965	0.961 to 0.970	−0.090	−0.153 to −0.026	MDL to 40
As	0.954	0.935 to 0.972	0.107	−0.091 to 0.305	0.3 to 45
B	1.294	1.287 to 1.301	−1.735	−6.905 to 3.434	3 to 2,000
Ba	0.891	0.832 to 0.949	−1.117	−5.782 to 3.548	31 to 203
Be	0.138	−0.102 to 0.378	−0.005	0.001 to 0.008	MDL to 0.06
Ca, mg/L	0.915	0.878 to 0.952	0.052	−2.047 to 2.151	13 to 150
Cd	0.926	0.924 to 0.928	0.002	−0.020 to 0.023	MDL to 47
Co	0.986	0.981 to 0.990	0.018	−0.049 to 0.085	MDL to 50
Cr	1.050	1.029 to 1.071	−0.003	−0.489 to 0.482	0.09 to 7,400
Cu	0.964	0.953 to 0.975	0.299	0.098 to 0.499	0.06 to 60
Fe	0.934	0.913 to 0.955	−0.035	−0.549 to 0.480	MDL to 110
K, mg/L	1.048	1.025 to 1.070	−0.112	−0.183 to −0.040	0.7 to 8.0
Li	1.208	1.160 to 1.255	5.659	1.708 to 9.610	0.2 to 230
Mg, mg/L	0.982	0.967 to 0.997	−0.779	−1.299 to −0.260	1.6 to 110
Mn	1.026	0.983 to 1.069	0.082	−1.256 to 1.420	MDL to 600
Mo	1.045	1.038 to 1.053	−0.112	−0.239 to 0.016	0.3 to 57
Na, mg/L	0.906	0.898 to 0.915	0.273	−0.263 to 0.810	1.6 to 220
Ni	1.045	1.035 to 1.056	0.494	0.285 to 0.704	0.1 to 70
Pb	0.922	0.917 to 0.928	0.024	−0.047 to 0.095	MDL to 42
Sb	0.950	0.949 to 0.952	0.016	0.006 to 0.026	MDL to 44
Se	0.929	0.913 to 0.944	0.134	−0.035 to 0.303	MDL to 52
SiO$_2$, mg/L	0.971	0.937 to 1.005	0.339	−0.890 to 1.567	3.9 to 74
Sr	0.870	0.829 to 0.911	68.799	9.040 to 128.559	30 to 5,000
Tl	0.873	0.870 to 0.875	0.011	−0.035 to 0.057	MDL to 58
U	0.920	0.905 to 0.935	0.246	0.089 to 0.403	0.08 to 45
V	1.010	1.005 to 1.014	−0.162	−0.803 to 0.478	0.2 to 750
W	0.874	0.873 to 0.876	0.031	−0.013 to 0.076	MDL to 132
Zn	1.020	1.013 to 1.028	−0.052	−0.692 to 0.587	0.4 to 280

Table 12. Linear regression results for unfiltered natural-water digestates analyzed using collision/reaction cell inductively coupled plasma–mass spectrometry and other spectrometric methods.

[Unfiltered natural-water digestates prepared using in-bottle digestion procedure (Hoffman and others, 1996) and filtered using 20-micrometer pore-size membrane; Analyte, concentrations are in micrograms per liter unless otherwise noted; Slope, see Appendix for other spectrometric method used in comparison; 95% CI, 95-percent confidence interval for the slope or y-intercept; y-inter, y intercept; Conc. range, concentration range of the 25 samples tested; MDL, method detection limit (see table 2); mg/L, milligrams per liter]

Analyte	Slope	95% CI for slope	y-inter	95% CI for y-inter	Conc. range
Ag	1.854	1.615 to 2.092	−0.011	−0.020 to −0.001	MDL to 0.12
Al	1.167	1.153 to 1.181	−21.799	−52.662 to 9.064	1.7 to 8,200
As	1.021	0.956 to 1.085	−0.413	−0.95 to 0.125	1.4 to 28
B	1.071	1.047 to 1.094	0.076	−1.160 to 1.312	3.8 to 140
Ba	0.970	0.951 to 0.990	−1.493	−3.517 to 0.531	4.6 to 365
Be	0.918	0.897 to 0.939	0.035	−0.009 to 0.080	MDL to 10
Ca, mg/L	1.000	0.954 to 1.046	1.193	−1.097 to 3.483	12 to 140
Cd	0.925	0.904 to 0.946	−0.002	−0.036 to 0.032	MDL to 7.9
Co	1.137	1.065 to 1.208	−0.127	−0.382 to 0.127	0.06 to 10
Cr	0.921	0.874 to 0.969	0.232	−0.274 to 0.738	0.4 to 44
Cu	1.220	1.147 to 1.294	−0.034	−0.463 to 0.396	0.2 to 15
Fe	1.078	1.069 to 1.087	−26.485	−44.006 to −8.963	5.6 to 10,000
K, mg/L	1.026	1.005 to 1.047	−0.210	−0.276 to −0.145	0.7 to 7.7
Li	0.863	0.821 to 0.906	1.472	0.018 to 2.926	MDL to 110
Mg, mg/L	0.993	0.975 to 1.011	−0.594	−1.206 to 0.019	1.5 to 104
Mn	0.938	0.907 to 0.969	1.682	−0.026 to 3.39	1.1 to 200
Mo	0.847	0.815 to 0.879	0.202	0.119 to 0.286	0.2 to 8.9
Na, mg/L	0.942	0.932 to 0.953	0.213	−0.382 to 0.807	1.6 to 196
Ni	1.128	1.080 to 1.176	0.188	0.016 to 0.360	0.3 to 17
Pb	1.089	1.043 to 1.135	0.039	−0.167 to 0.244	0.04 to 17
Sb	1.846	1.484 to 2.207	−0.161	−0.249 to −0.073	0.06 to 0.63
Se	0.831	0.695 to 0.967	0.255	0.133 to 0.378	0.08 to 2.1
SiO$_2$, mg/L	1.136	1.014 to 1.259	0.238	−1.999 to 2.475	3.3 to 37
Sr	1.004	0.987 to 1.020	−2.933	−6.698 to 0.831	27 to 720
Tl	0.942	0.874 to 1.009	0.072	0.031 to 0.113	MDL to 3.0
U	1.024	0.951 to 1.097	0.046	−0.313 to 0.405	0.2 to 14
V	0.900	0.862 to 0.939	0.716	−0.035 to 1.466	1.4 to 75
W	1.122	0.654 to 1.589	−0.016	−0.039 to 0.008	MDL to 0.13
Zn	0.995	0.983 to 1.008	0.538	−0.560 to 1.637	0.7 to 340

Table 13. Bias and variability of results for digested biological standard reference materials analyzed by collision/reaction cell inductively coupled plasma–mass spectrometry.

[Analyte, concentrations are in micrograms per gram unless otherwise noted; Dolt-2, National Research Council Canada (NRC) dogfish liver standard reference material; Dorm-2, NRC dogfish muscle standard reference material; Bovine, National Institute of Standards and Technology bovine liver standard reference material 1577b; Exp, mean concentration plus or minus one standard deviation for experimental results based on three replicate measurements; Target, certified concentration (plus or minus one standard deviation when available), numbers in parentheses are uncertified; mg/g, milligrams per gram; na, concentration not available]

Analyte	Dolt-2		Dorm-2		Bovine	
	Exp	Target	Exp	Target	Exp	Target
Ag	0.655±0.005	0.608±0.032	0.035±0.007	0.041±0.013	0.043±0.009	0.039
Al	36±1	25.2±2.4	12.4±0.5	10.9±1.7	1.06±0.05	(3)
As	18.31±0.06	16.6±1.1	20.7±0.1	18.0±1.1	0.071±0.008	(0.05)
Ca, mg/g	na	na	na	na	0.13±0.01	0.116
Cd	24.4±0.3	20.8±0.5	0.07±0.2	0.043±0.008	0.6±0.2	0.50
Co	0.24±0.02	0.24±0.05	0.22±0.02	0.182±0.031	0.27±0.02	(0.25)
Cr	0.30±0.01	0.37±0.08	37.4±0.6	34.7±5.5	na	na
Cu	29±2	25.8±1.1	2.1±0.1	2.34±0.16	160±10	160
Fe	1,290±5	1,103±47	183.6±0.8	142±10	217±2	184
K, mg/g	na	na	na	na	10.54±0.04	9.94
Mg, mg/g	na	na	na	na	0.67±0.01	0.601
Mn	7.2±0.3	6.88±0.56	4.3±0.2	3.66±0.34	11.7±0.5	10.5
Mo	na	na	na	na	4.1±0.2	3.5
Na, mg/g	na	na	na	na	2.49±0.09	2.42
Ni	0.17±0.02	0.20±0.02	19.6±0.4	19.4±3.1	na	na
Pb	0.25±0.02	0.22±0.02	0.045±0.004	0.065±0.007	0.124±0.008	0.129
Se	7.7±0.1	6.06±0.49	1.77±0.06	1.4±0.09	0.940±0.009	0.73
Sr	na	na	na	na	0.143±0.003	0.136
V	na	na	na	na	0.109±0.001	(0.123)
Zn	104±3	85.8±2.5	26.7±0.6	25.6±2.3	140±4	127

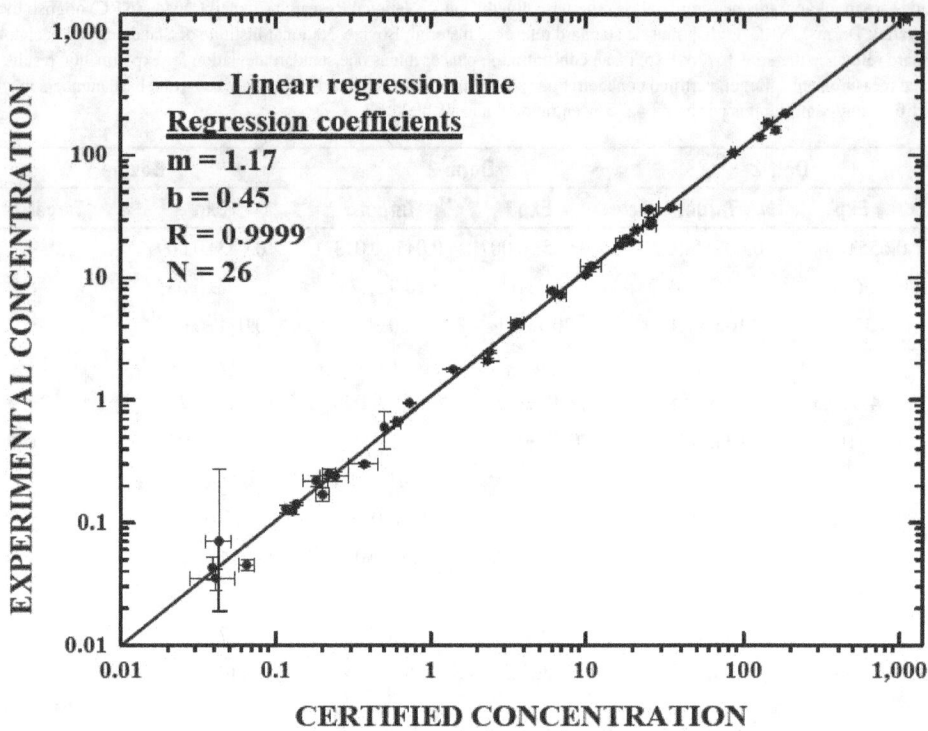

Figure 6. Linear regression analysis of experimental concentrations for all elements in biological standard reference material digestates analyzed by collision/reaction cell inductively coupled plasma–mass spectrometry in relation to the certified concentrations. Error bars in each axis represent one standard deviation. The regression coefficients are the slope (m), *y*-intercept (b), correlation coefficient (R), and the number of data points (N). Concentrations are in micrograms per gram or milligrams per gram, dry weight. The dataset is listed in table 13.

Summary and Conclusions

A new analytical method using collision/reaction cell inductively coupled plasma mass spectrometry (cICP–MS) has been shown to eliminate molecular ion interferences associated with selected elements. Helium or hydrogen was used to eliminate or substantially reduce such interferences. The scope of the updated method was expanded by including elements that were not determined by earlier inductively coupled plasma–mass spectrometric (ICP–MS) methods. The updated method can be used to determine elements in filtered natural water and in a wide variety of filtered aqueous matrices including whole water, biota, sediment, and soil digestates. Elements that are not affected by molecular ion interference also can be determined with bias and variability comparable to earlier ICP–MS methods.

The findings of this report are summarized as follows to address previously cited objectives:

- Method detection limits were established for all elements determined using the new method. Most method detection limits are lower or unchanged compared to earlier methods except for Co, K, Mg, Ni, SiO_2, and Tl, which are less than a factor of 2 higher.

- The results presented have shown that As, Ca, Co, Cr,Cu, Fe, K, Mg, Na, Ni, Se, Si, V, W, and Zn can be determined accurately when using a collision/reaction gas without mathematical corrections in the presence of interferent concentrations substantially greater than those found in routinely submitted samples. A collision/reaction gas was not needed when determining Ag, Al, Cd, and Mn because the interferences were insignificant at the interferent concentrations tested.

- Samples having specific conductances less than 7,500 µS/cm can be analyzed directly by cICP–MS with acceptable bias and variability. Previous ICP–MS methods required that samples with specific conductances greater than 2,500 µS/cm be diluted. Ninety-five percent of the samples submitted to NWQL in 2004 had specific conductances less than 3,000 µS/cm.

- The bias and variability of the cICP–MS method were determined using results from aqueous standard reference materials. The average results for the two standard reference materials tested were within one F-pseudosigma of the most probable concentration except for Ba, Na, Mg, Ni, and U. Nevertheless, the averages for these elements were within an acceptable 1.5 F-pseudosigma of the most probable concentration. Elemental results for biological standard reference materials also compared closely to the certified concentrations.

- The new cICP–MS method was validated by using analytical results from reagent-water, ground-water, and surface-water matrices that were spiked at 2 to 4 concentration levels. The percentage biases for reagent water, ground water, and surface water ranged from –14 to 21 percent for trace elements and –7.7 to 10 percent for Ca, K, Mg, Na, and SiO_2.

- The cICP–MS results from 25 filtered natural-water samples and 25 whole-water digestates were compared with results from previously used methods. Furthermore, experimental results for three biological standard reference material digestates were compared to certified concentrations. Linear regression analysis of these results indicated that the bias and variability of cICP–MS is acceptable for a wide variety of sample matrices.

References Cited

Agilent Technologies, 2001, Agilent 1100 series reference manuals for micro autosampler, quaternary pump, thermostatted column compartment, and thermostatted autosampler: Part numbers G1313-90004, G1316-90003, G1311-90003, and G1330-90002, Waldbronn, Germany.

Agilent Technologies, 2004a, Agilent 7500 ICP–MS Chem-Station operator's manual: Part number G3270-90110010, Rev. A 2004/04, Tokyo, Japan.

Agilent Technologies, 2004b, Agilent 7500 ICP–MS hardware manual: Part number G3270-90100, Rev. A 2004/04, Tokyo, Japan.

American Society for Testing and Materials, 1999, E29-93a Standard practice for using significant digits in test data to determine conformance to specifications: Annual book of ASTM standards, Reporting test results, section 7.4, v. 14.02.

American Society for Testing and Materials, 2000, Annual book of ASTM standards, Section 11, Water: Philadelphia, American Society for Testing and Materials, v. 11.01, D1193, p. 10.

Childress, C.J.O., Foreman, W.T., Connor, B.F., and Maloney, T.J., 1999, New reporting procedures based on long-term method detection levels and some considerations for interpretations of water-quality data provided by the U.S. Geological Survey National Water Quality Laboratory: U.S. Geological Survey Open-File Report 99-193, 19 p.

Crawford, J.K., and Luoma, S.N., 1993, Guidelines for studies of contaminants in biological tissues for the National Water-Quality Assessment Program: U.S. Geological Survey Open-File Report 92-494, 69 p.

Faires, L.M., 1993, Methods of analysis by the U.S. Geological Survey National Water Quality Laboratory—Determination of metals in water by inductively coupled plasma–mass spectrometry: U.S. Geological Survey Open-File Report 92-634, 28 p.

Farrar, J.W., and Long, H.K., 1997, Report on the U.S. Geological Survey's evaluation program for standard reference samples distributed in September 1996: U.S. Geological Survey Open-File Report 97-20, 145 p.

Fishman, M.J., and Friedman, L.C., eds., 1989, Methods of determination of inorganic substances in water and fluvial sediments: U.S. Geological Survey Techniques of Water-Resources Investigations, book 5, chap. A1, 545 p.

Garbarino, J.R., and Taylor, H.E., 1994, Inductively coupled plasma–mass spectrometric method for the determination of dissolved trace elements in natural water: U.S. Geological Survey Open-File Report 94-358, 28 p.

Garbarino, J.R., and Struzeski, T.M., 1998, Methods of analysis by the U.S. Geological Survey National Water Quality Laboratory—Determination of elements in whole-water digests using inductively coupled plasma–optical emission spectrometry and inductively coupled plasma–mass spectrometry: U.S. Geological Survey Open-File Report 98-165, 101 p.

Garbarino, J.R. 1999, Methods of analysis by the U.S. Geological Survey National Water Quality Laboratory—Determination of dissolved arsenic, boron, lithium, selenium, strontium, thallium, and vanadium using inductively coupled plasma–mass spectrometry: U.S. Geological Survey Open-File Report 99-093, 31 p.

Garbarino, J.R., 2000, Methods of analysis by the U.S. Geological Survey National Water Quality Laboratory—Determination of whole-water recoverable arsenic, boron, and vanadium using inductively coupled plasma–mass spectrometry: U.S. Geological Survey Open-File Report 99-464, 15 p.

Garbarino, J.R., Bednar, A.J., and Burkhardt, M.R., 2002, Methods of analysis by the U.S. Geological Survey National Water Quality Laboratory—Arsenic speciation in natural-water samples using laboratory and field methods: U.S. Geological Survey Water-Resources Investigations Report 02-4144, 40 p.

Hansen, W.R., 1991, Suggestions to authors of the reports of the United States Geological Survey (7th ed.): U.S. Government Printing Office, p. 119–121.

Hem, J.D., 1986, Study and interpretation of the chemical characteristics of natural water: U.S. Geological Survey Water-Supply Paper 2254, p. 67.

Hoaglin, D.C., 1983, Letter values: A set of selected order statistics, in Hoaglin, D.C., Mosteller, Frederick, and Tukey, J.W., eds., Understanding robust and exploratory data analysis: New York, John Wiley & Sons, Inc., chap. 2, p. 35–55.

Hoffman, G.L., Fishman, M.J., and Garbarino, J.R., 1996, Methods of analysis by the U.S. Geological Survey National Water Quality Laboratory—In-bottle acid digestion of whole-water samples: U.S. Geological Survey Open-File Report 96-225, 28 p.

Horlick, Gary, and Montaser, Akbar, 1998, Analytical characteristics of ICPMS, in Montaser, Akbar, ed., Inductively coupled plasma mass spectrometry: New York, Wiley-VCH, p. 503–613.

Maloney, T.J., 2005, Quality management system, U.S. Geological Survey National Water Quality Laboratory: U.S. Geological Survey Open-File Report 2005-1263, version 1.3, November 9, 2005, accessed January 25, 2006, at http://pubs.water.usgs.gov/ofr2005-1263/

National Environmental Laboratory Accreditation Conference, 2003, Glossary for NELAC: 2003 NELAC Standard, EPA/600/R-04/003, accessed September 8, 2005, at URL http://www.epa.gov/nelac/2003standards.html.

Novak, C.E., 1985, Preparation of water-resources data reports: U.S. Geological Survey Open-File Report 85-480, 331 p.

Tan, S.H., and Horlick, G., 1986, Background spectral features in inductively coupled plasma/mass spectrometry: Applied Spectroscopy, v. 40, p. 445–460.

Taylor, J.K., 1990, Statistical techniques for data analysis: Lewis Publishers, Inc., CRC Press, Boca Raton, Florida, ISBN 0-87371-250-1.

U.S. Environmental Protection Agency, 1996, Microwave assisted acid digestion of siliceous and organically based matrices: U.S. Environmental Protection Agency Method 3052, Revision 0, December 1996, 20 p.

U.S. Environmental Protection Agency, 1998, Microwave assisted acid digestion of sediments, sludges, soils, and oils: U.S. Environmental Protection Agency Method 3051A, Revision 1, January 1998, 25 p.

U.S. Environmental Protection Agency, 2000, Guideline establishing test procedures for the analysis of pollutants (Part 136, Appendix B. Definition and procedure for the determination of the method detection limit—Revision 1.11): U.S. Code of Federal Regulations, Title 40, revised as of July 1, 2000.

U.S. Environmental Protection Agency, 2005, Test methods for evaluating solid waste, physical/chemical methods: U.S. Environmental Protection Agency SW-846 manual (3rd ed.), publication number 955-001-00000-1, accessed September 6, 2005, at URL http://www.epa.gov/epaoswer/hazwaste/test/main.htm.

U.S. Geological Survey, 2002, Policy for storing and reporting significant figures for chemical data: Office of Water Quality Technical Memorandum No. 2002.11, accessed September 6, 2005, at URL http://water.usgs.gov/admin/memo/QW/qw02.11.html.

U.S. Geological Survey, variously dated, National field manual for the collection of water-quality data: U.S. Geological Survey Techniques of Water-Resources Investigations, book 9, chaps. A1 (Rev. 1/2005), A2 (Rev. 3/2003), A3 (Rev. 4/2004), A4 (Rev. 9/1999), A5 (Rev. 4/2004), and A5.6.4 (Rev. 1/2005), accessed September 6, 2005, at URL http://water.usgs.gov/owq/FieldManual/.

Woodworth, M.T., and Conner, B.F., 2002, Results of the U.S. Geological Survey's analytical evaluation program for standard reference samples distributed in March 2002: U.S. Geological Survey Open-File Report 02-243, 113 p.

Appendix

Linear Regression Analyses of Elemental Results

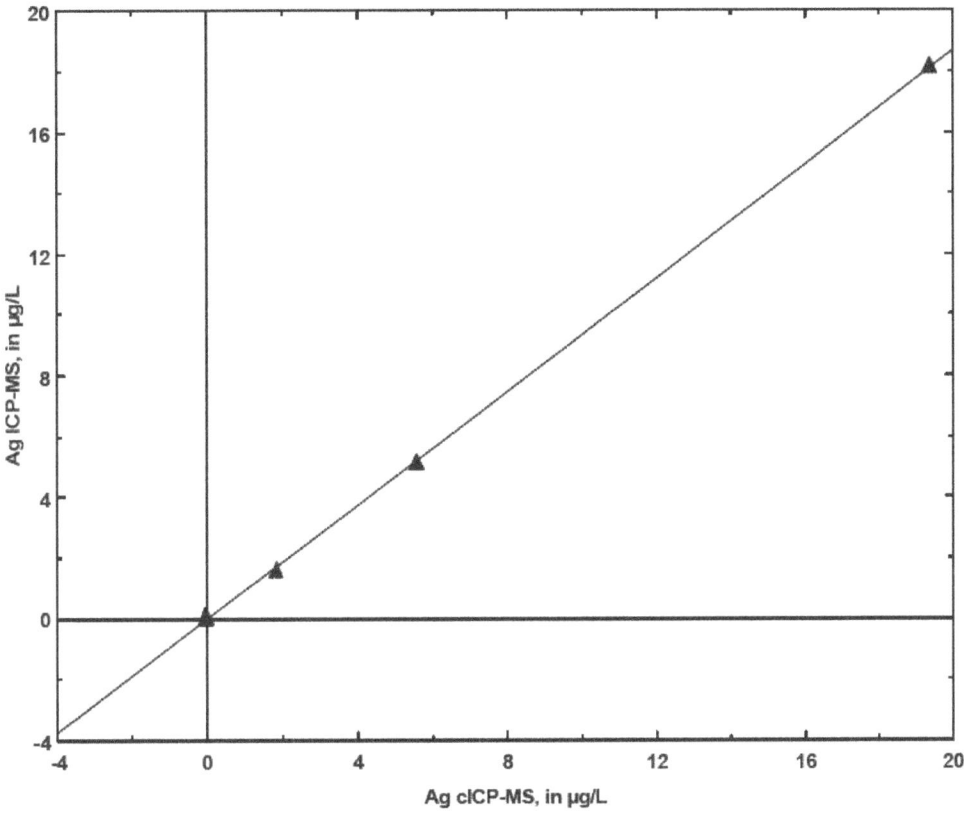

LINEAR REGRESSION EQUATION

Ag ICP-MS, in µg/L = -.006 + .933 * Ag cICP-MS, in µg/L; R^2 = 1

Confidence Intervals
Ag ICP-MS, in µg/L vs. Ag cICP-MS, in µg/L

	Coefficient	95% Lower	95% Upper
Intercept	-.006	-.022	.009
Ag cICP-MS, in µg/L	.933	.929	.936

Figure A1. Linear regression analysis of silver results from filtered water samples analyzed by inductively coupled plasma–mass spectrometry (ICP–MS) and collision/reaction cell inductively coupled plasma–mass spectrometry (cICP–MS). R^2 is the coefficient of determination. In the confidence intervals table, the Intercept and Ag cICP–MS coefficients are the *y*-intercept and slope, respectively. Results are in micrograms per liter (µg/L).

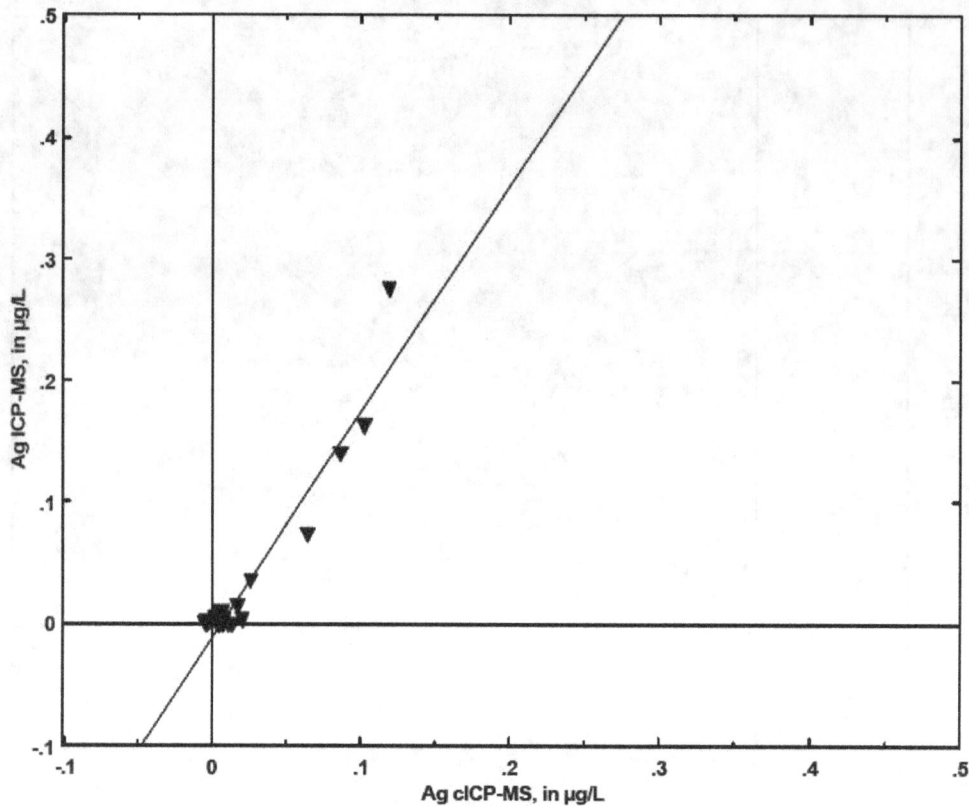

LINEAR REGRESSION EQUATION

Ag ICP-MS, in µg/L = -.011 + 1.854 * Ag cICP-MS, in µg/L; R^2 = .918

Confidence Intervals
Ag ICP-MS, in µg/L vs. Ag cICP-MS, in µg/L

	Coefficient	95% Lower	95% Upper
Intercept	-.011	-.020	-.001
Ag cICP-MS, in µg/L	1.854	1.615	2.092

Figure A2. Linear regression analysis of silver results from whole-water recoverable digestates analyzed by inductively coupled plasma–mass spectrometry (ICP–MS) and collision/reaction cell inductively coupled plasma–mass spectrometry (cICP–MS). R^2 is the coefficient of determination. In the confidence intervals table, the Intercept and Ag cICP–MS coefficients are the *y*-intercept and slope, respectively. Results are in micrograms per liter (µg/L).

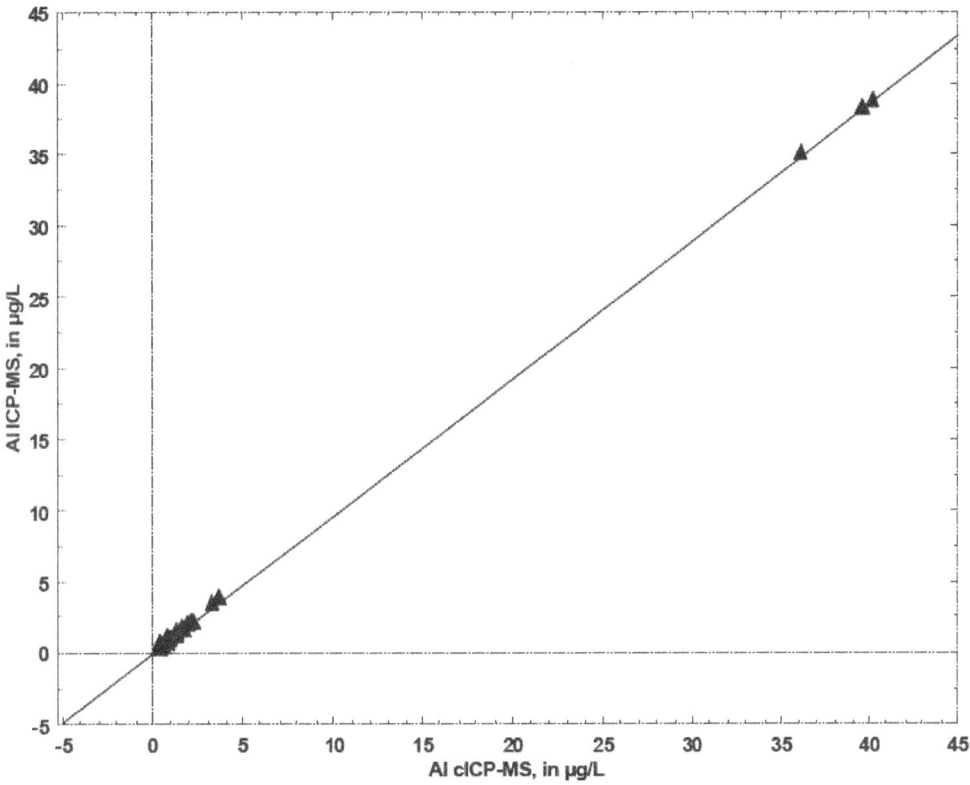

LINEAR REGRESSION EQUATION

Al ICP-MS, in µg/L = -.09 + .965 * Al cICP-MS, in µg/L; R^2 = 1

Confidence Intervals
Al ICP-MS, in µg/L vs. Al cICP-MS, in µg/L

	Coefficient	95% Lower	95% Upper
Intercept	-.090	-.153	-.026
Al cICP-MS, in µg/L	.965	.961	.970

Figure A3. Linear regression analysis of aluminum results from filtered water samples analyzed by inductively coupled plasma–mass spectrometry (ICP–MS) and collision/reaction cell inductively coupled plasma–mass spectrometry (cICP–MS). R^2 is the coefficient of determination. In the confidence intervals table, the Intercept and Al cICP–MS coefficients are the y-intercept and slope, respectively. Results are in micrograms per liter (µg/L).

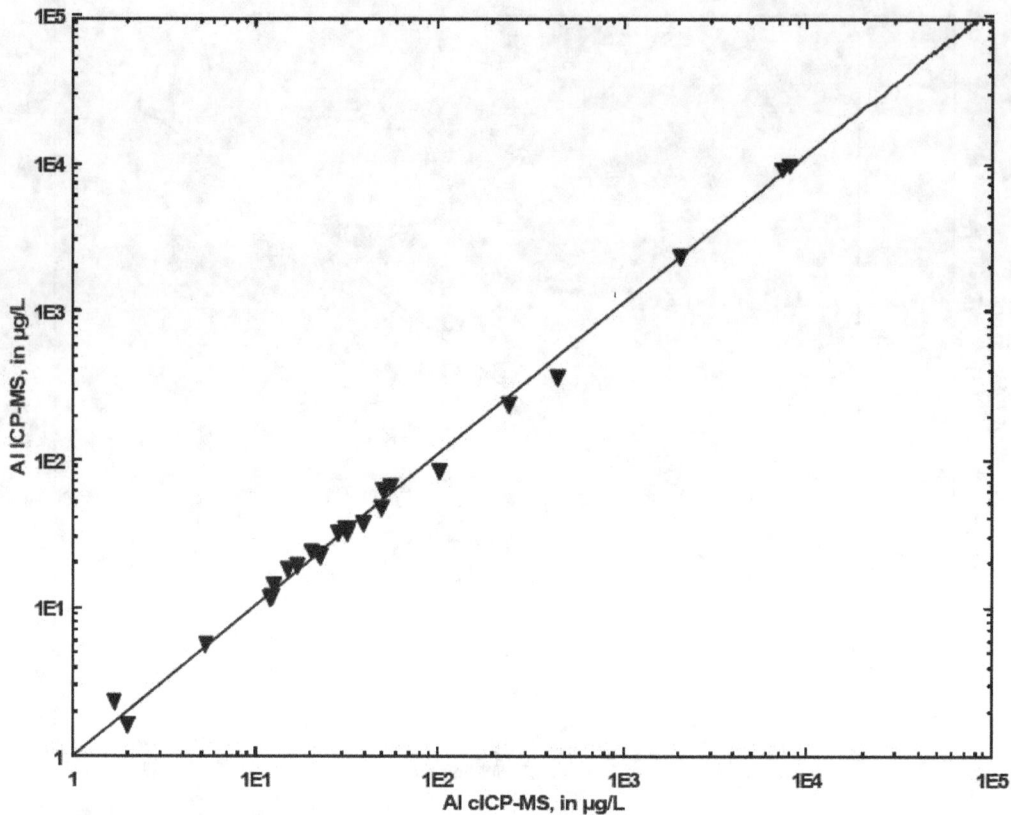

LINEAR REGRESSION EQUATION

Al ICP-MS, in µg/L = -21.799 + 1.167 * Al cICP-MS, in µg/L; R^2 = .999

Confidence Intervals
Al ICP-MS, in µg/L vs. Al cICP-MS, in µg/L

	Coefficient	95% Lower	95% Upper
Intercept	-21.799	-52.662	9.064
Al cICP-MS, in µg/L	1.167	1.153	1.181

Figure A4. Linear regression analysis of aluminum results from whole-water recoverable digestates analyzed by inductively coupled plasma–mass spectrometry (ICP–MS) and collision/reaction cell inductively coupled plasma–mass spectrometry (cICP–MS). R^2 is the coefficient of determination. In the confidence intervals table, the Intercept and Al cICP–MS coefficients are the y-intercept and slope, respectively. Results are in micrograms per liter (µg/L).

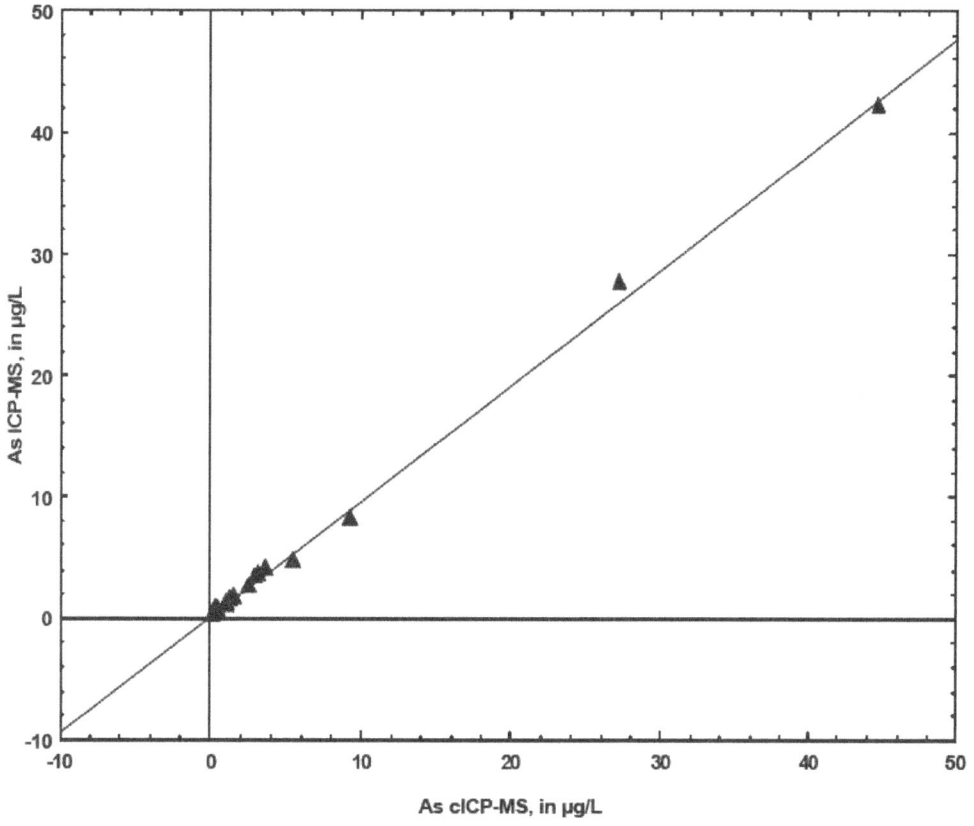

LINEAR REGRESSION EQUATION

As ICP-MS, in µg/L = .107 + .954 * As cICP-MS, in µg/L; R^2 = .998

Confidence Intervals
As ICP-MS, in µg/L vs. As cICP-MS, in µg/L

	Coefficient	95% Lower	95% Upper
Intercept	.107	-.091	.305
As cICP-MS, in µg/L	.954	.935	.972

Figure A5. Linear regression analysis of arsenic results from filtered water samples analyzed by inductively coupled plasma–mass spectrometry (ICP–MS) and collision/reaction cell inductively coupled plasma–mass spectrometry (cICP–MS). R^2 is the coefficient of determination. In the confidence intervals table, the Intercept and As cICP–MS coefficients are the y-intercept and slope, respectively. Results are in micrograms per liter (µg/L).

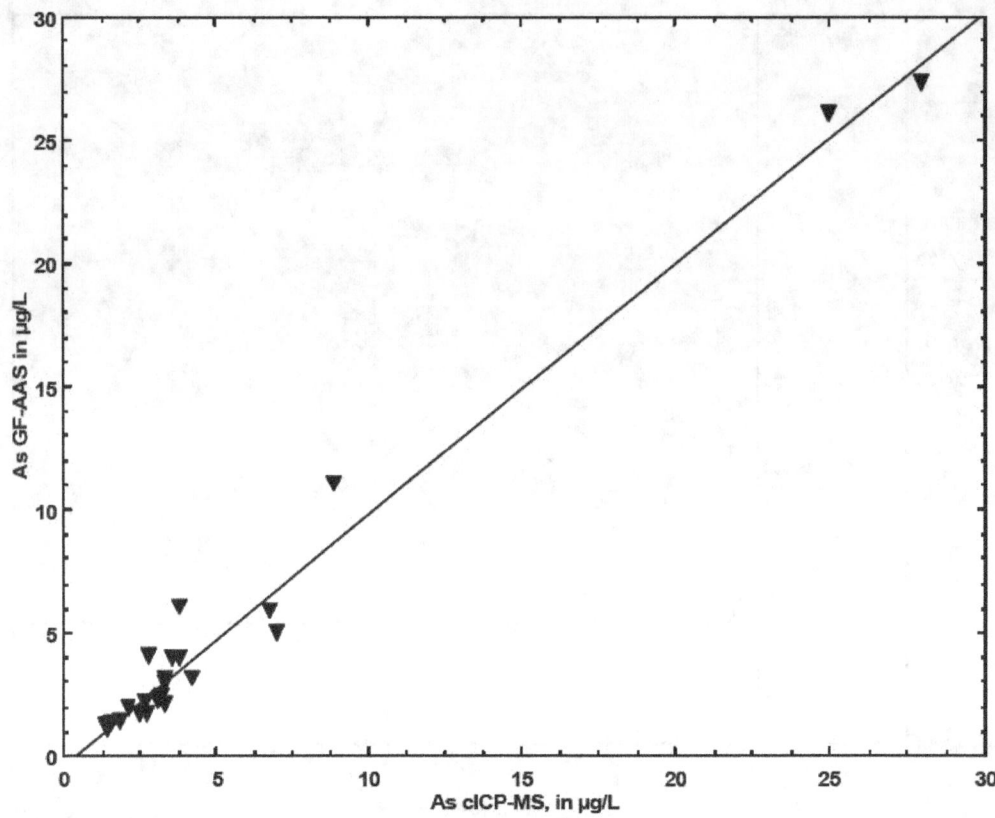

LINEAR REGRESSION EQUATION

As GF-AAS in µg/L = -.413 + 1.021 * As cICP-MS, in µg/L; R^2 = .979

Confidence Intervals
As GF-AAS in µg/L vs. As cICP-MS, in µg/L

	Coefficient	95% Lower	95% Upper
Intercept	-.413	-.950	.125
As cICP-MS, in µg/L	1.021	.956	1.085

Figure A6. Linear regression analysis of arsenic results from whole-water recoverable digestates analyzed by graphite furnace–atomic absorption spectrometry (GF–AAS) and collision/reaction cell inductively coupled plasma–mass spectrometry (cICP–MS). R^2 is the coefficient of determination. In the confidence intervals table, the Intercept and As cICP–MS coefficients are the *y*-intercept and slope, respectively. Results are in micrograms per liter (µg/L).

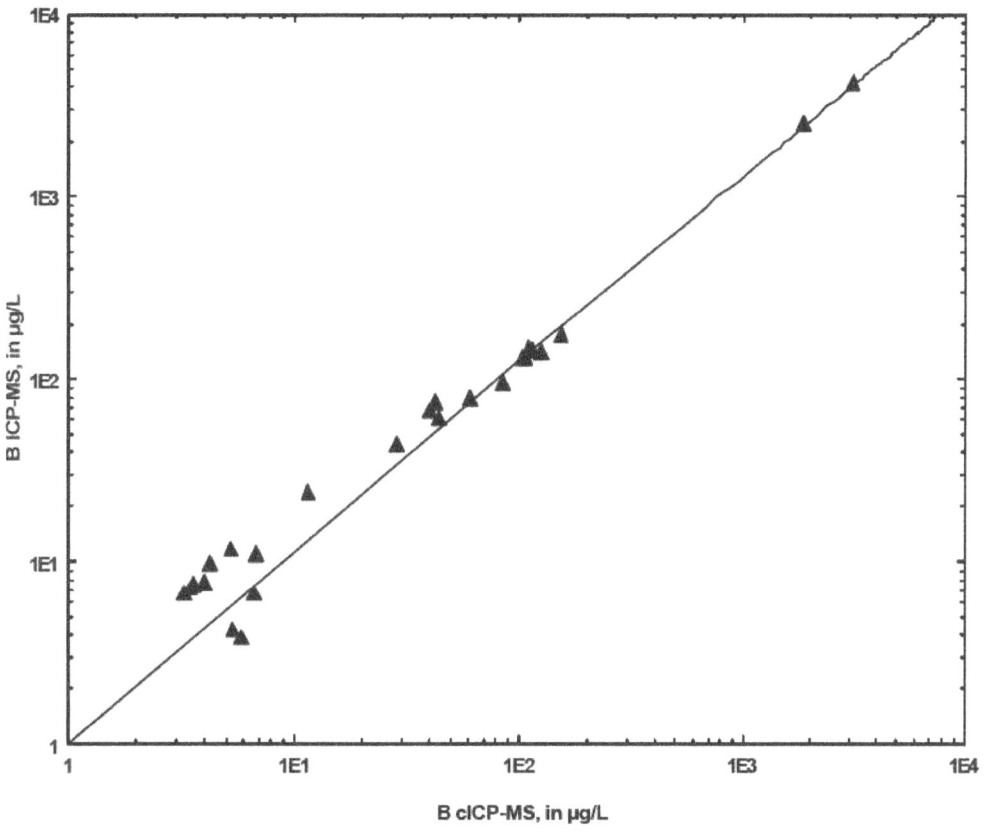

LINEAR REGRESSION EQUATION

LINEAR REGRESSION EQUATION

B ICP-MS, in µg/L = -1.735 + 1.294 * B cICP-MS, in µg/L; R^2 = 1

Confidence Intervals
B ICP-MS, in µg/L vs. B cICP-MS, in µg/L

	Coefficient	95% Lower	95% Upper
Intercept	-1.735	-6.905	3.434
B cICP-MS, in µg/L	1.294	1.287	1.301

Figure A7. Linear regression analysis of boron results from filtered water samples analyzed by inductively coupled plasma–mass spectrometry (ICP–MS) and collision/reaction cell inductively coupled plasma–mass spectrometry (cICP–MS). R^2 is the coefficient of determination. In the confidence intervals table, the Intercept and B cICP–MS coefficients are the y-intercept and slope, respectively. Results are in micrograms per liter (µg/L).

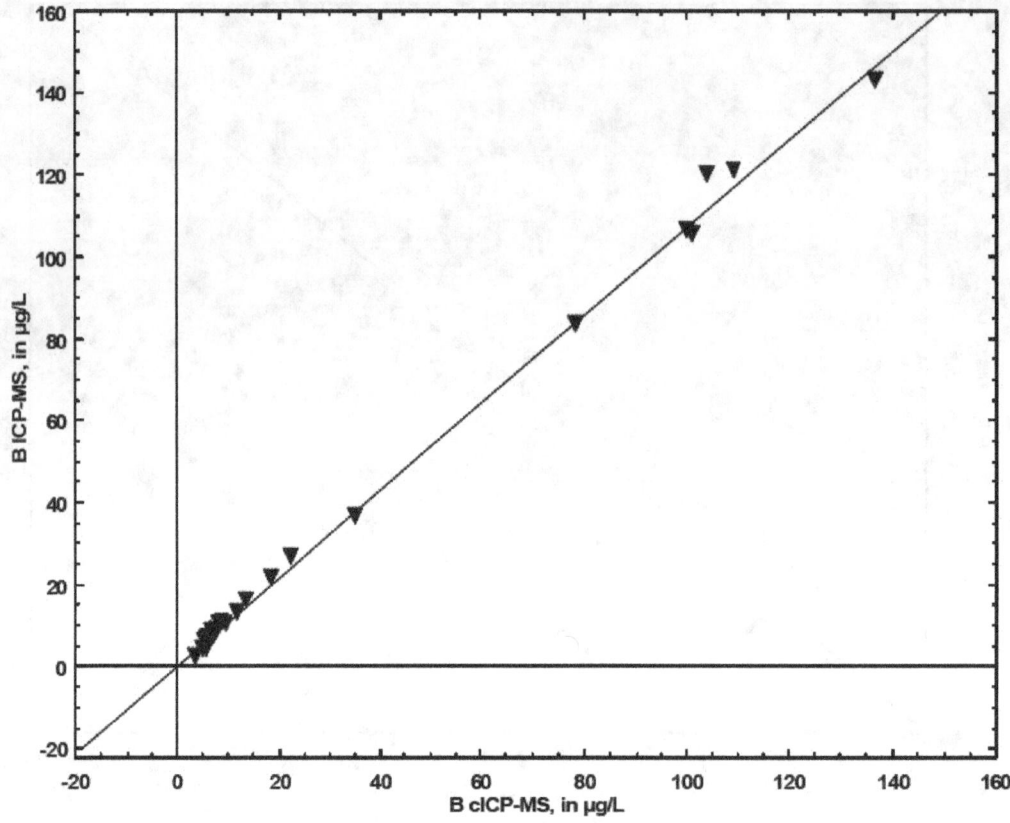

LINEAR REGRESSION EQUATION

B ICP-MS, in µg/L = .076 + 1.071 * B cICP-MS, in µg/L; R^2 = .997

Confidence Intervals
B ICP-MS, in µg/L vs. B cICP-MS, in µg/L

	Coefficient	95% Lower	95% Upper
Intercept	.076	-1.160	1.312
B cICP-MS, in µg/L	1.071	1.047	1.094

Figure A8. Linear regression analysis of boron results from whole-water recoverable digestates analyzed by inductively coupled plasma–mass spectrometry (ICP–MS) and collision/reaction cell inductively coupled plasma–mass spectrometry (cICP–MS). R^2 is the coefficient of determination. In the confidence intervals table, the Intercept and B cICP–MS coefficients are the y-intercept and slope, respectively. Results are in micrograms per liter (µg/L).

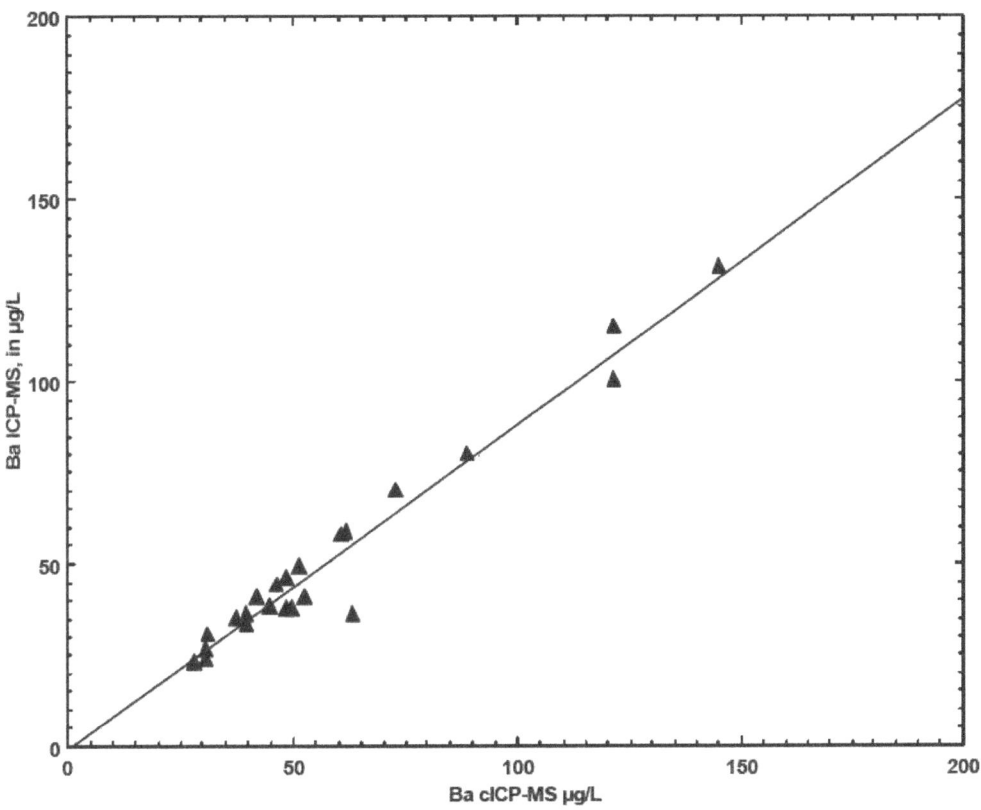

LINEAR REGRESSION EQUATION

Ba ICP-MS, in µg/L = -1.117 + .891 * Ba cICP-MS µg/L; R^2 = .977

Confidence Intervals
Ba ICP-MS, in µg/L vs. Ba cICP-MS µg/L

	Coefficient	95% Lower	95% Upper
Intercept	-1.117	-5.782	3.548
Ba cICP-MS µg/L	.891	.832	.949

Figure A9. Linear regression analysis of barium results from filtered water samples analyzed by inductively coupled plasma–mass spectrometry (ICP–MS) and collision/reaction cell inductively coupled plasma–mass spectrometry (cICP–MS). R^2 is the coefficient of determination. In the confidence intervals table, the Intercept and Ba cICP–MS coefficients are the *y*-intercept and slope, respectively. Results are in micrograms per liter (µg/L).

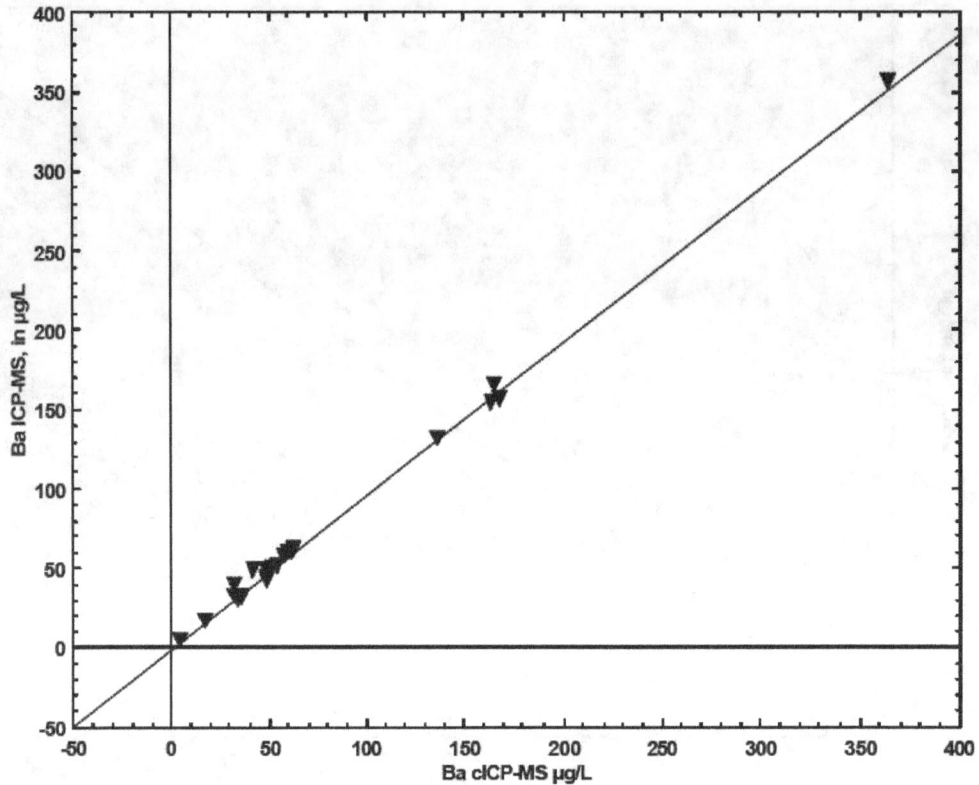

LINEAR REGRESSION EQUATION

Ba ICP-MS, in µg/L = -1.493 + .97 * Ba cICP-MS µg/L; R^2 = .998

Confidence Intervals
Ba ICP-MS, in µg/L vs. Ba cICP-MS µg/L

	Coefficient	95% Lower	95% Upper
Intercept	-1.493	-3.517	.531
Ba cICP-MS µg/L	.970	.951	.990

Figure A10. Linear regression analysis of barium results from whole-water recoverable digestates analyzed by inductively coupled plasma–mass spectrometry (ICP–MS) and collision/reaction cell inductively coupled plasma–mass spectrometry (cICP–MS). R^2 is the coefficient of determination. In the confidence intervals table, the Intercept and Ba cICP–MS coefficients are the y-intercept and slope, respectively. Results are in micrograms per liter (µg/L).

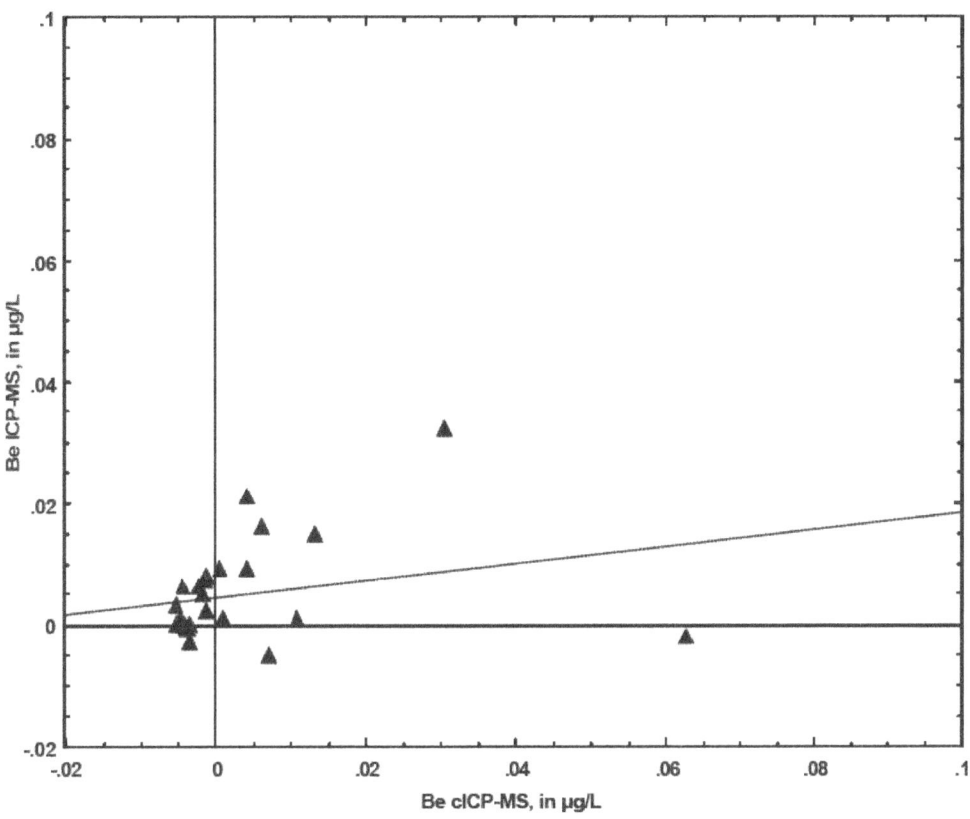

LINEAR REGRESSION EQUATION

Be ICP-MS, in µg/L = .005 + .138 * Be cICP-MS, in µg/L; R^2 = .058

Confidence Intervals
Be ICP-MS, in µg/L vs. Be cICP-MS, in µg/L

	Coefficient	95% Lower	95% Upper
Intercept	.005	.001	.008
Be cICP-MS, in µg/L	.138	-.102	.378

Figure A11. Linear regression analysis of beryllium results from filtered water samples analyzed by inductively coupled plasma–mass spectrometry (ICP–MS) and collision/reaction cell inductively coupled plasma–mass spectrometry (cICP–MS). R^2 is the coefficient of determination. In the confidence intervals table, the Intercept and Be cICP–MS coefficients are the y-intercept and slope, respectively. Results are in micrograms per liter (µg/L).

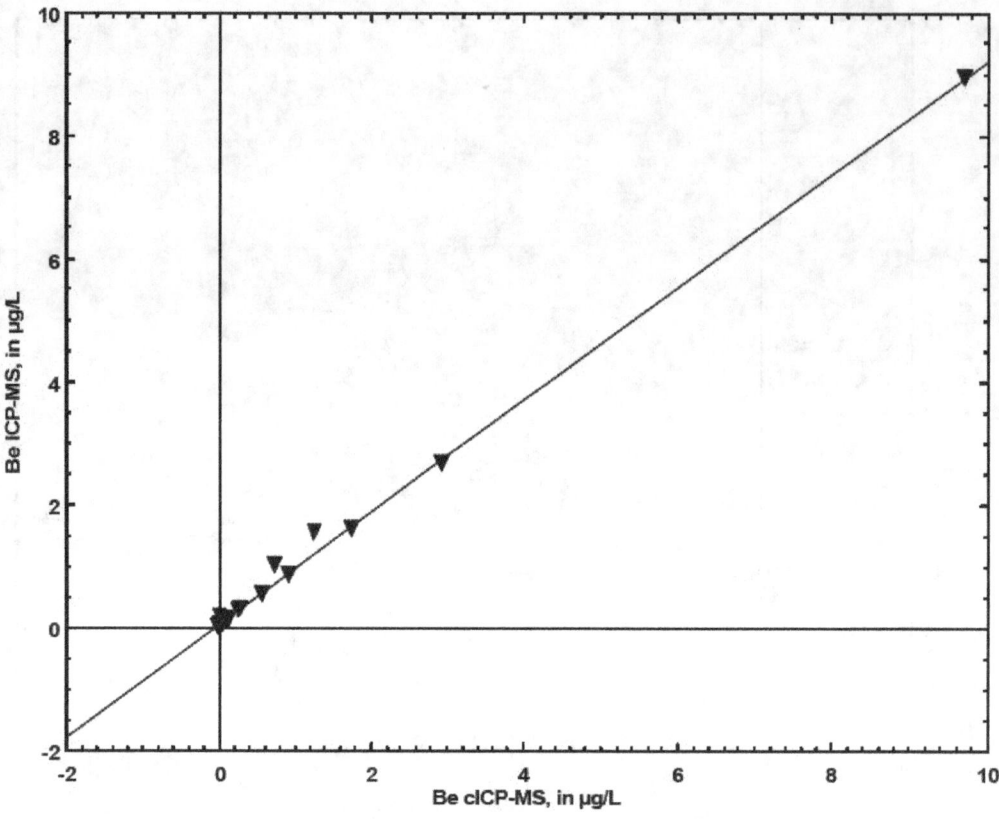

LINEAR REGRESSION EQUATION

Be ICP-MS, in µg/L = .035 + .918 * Be cICP-MS, in µg/L; R^2 = .997

Confidence Intervals
Be ICP-MS, in µg/L vs. Be cICP-MS, in µg/L

	Coefficient	95% Lower	95% Upper
Intercept	.035	-.009	.080
Be cICP-MS, in µg/L	.918	.897	.939

Figure A12. Linear regression analysis of beryllium results from whole-water recoverable digestates analyzed by inductively coupled plasma–mass spectrometry (ICP–MS) and collision/reaction cell inductively coupled plasma–mass spectrometry (cICP–MS). R^2 is the coefficient of determination. In the confidence intervals table, the Intercept and Be cICP–MS coefficients are the y-intercept and slope, respectively. Results are in micrograms per liter (µg/L).

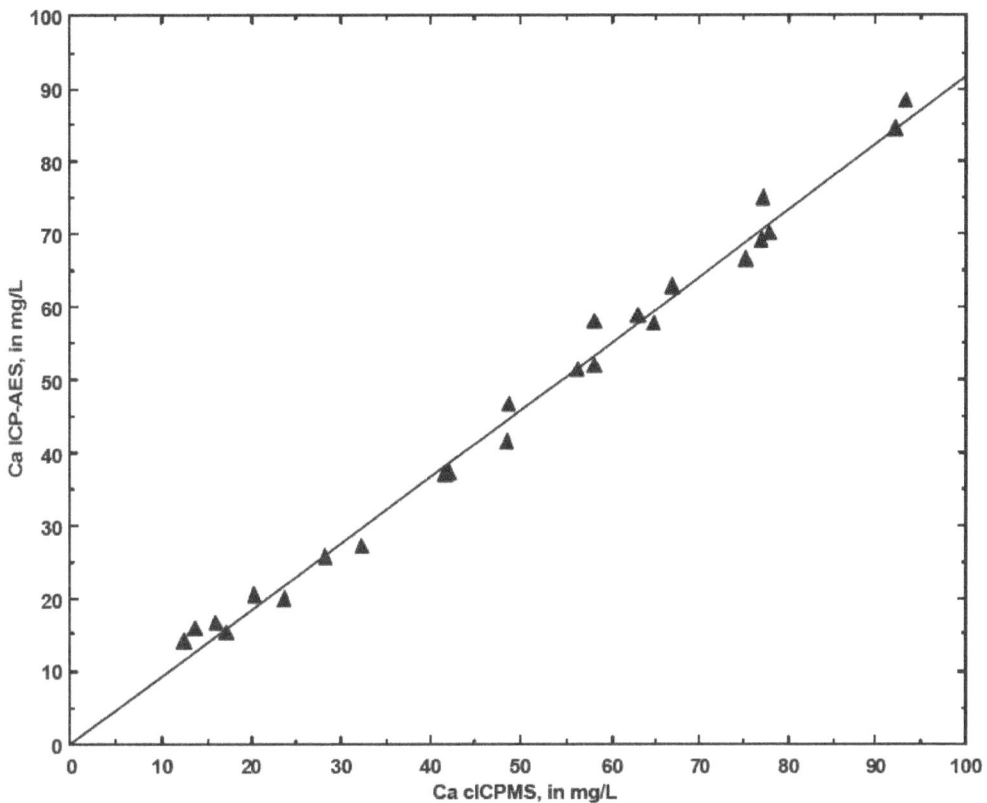

LINEAR REGRESSION EQUATION

Ca ICP-AES, in mg/L = .052 + .915 * Ca cICPMS, in mg/L; R^2 = .991

Confidence Intervals
Ca ICP-AES, in mg/L vs. Ca cICPMS, in mg/L

	Coefficient	95% Lower	95% Upper
Intercept	.052	-2.047	2.151
Ca cICPMS, in mg/L	.915	.878	.952

Figure A13. Linear regression analysis of calcium results from filtered water samples analyzed by inductively coupled plasma–atomic emission spectrometry (ICP–AES) and collision/reaction cell inductively coupled plasma–mass spectrometry (cICP–MS). R^2 is the coefficient of determination. In the confidence intervals table, the Intercept and Ca cICP–MS coefficients are the y-intercept and slope, respectively. Results are in milligrams per liter (mg/L).

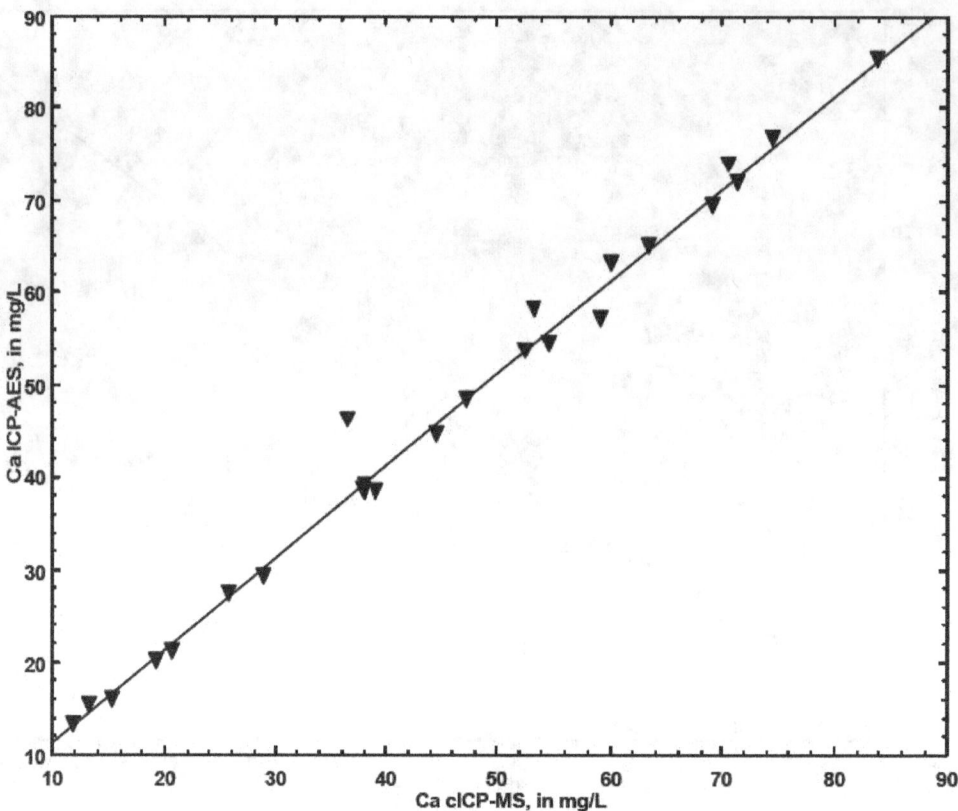

LINEAR REGRESSION EQUATION

Ca ICP-AES, in mg/L = 1.193 + 1 * Ca cICP-MS, in mg/L; R^2 = .989

Confidence Intervals
Ca ICP-AES, in mg/L vs. Ca cICP-MS, in mg/L

	Coefficient	95% Lower	95% Upper
Intercept	1.193	-1.097	3.483
Ca cICP-MS, in mg/L	1.000	.954	1.046

Figure A14. Linear regression analysis of calcium results from whole-water recoverable digestates analyzed by inductively coupled plasma–atomic emission spectrometry (ICP–AES) and collision/reaction cell inductively coupled plasma–mass spectrometry (cICP–MS). R^2 is the coefficient of determination. In the confidence intervals table, the Intercept and Ca cICP–MS coefficients are the *y*-intercept and slope, respectively. Results are in milligrams per liter (mg/L).

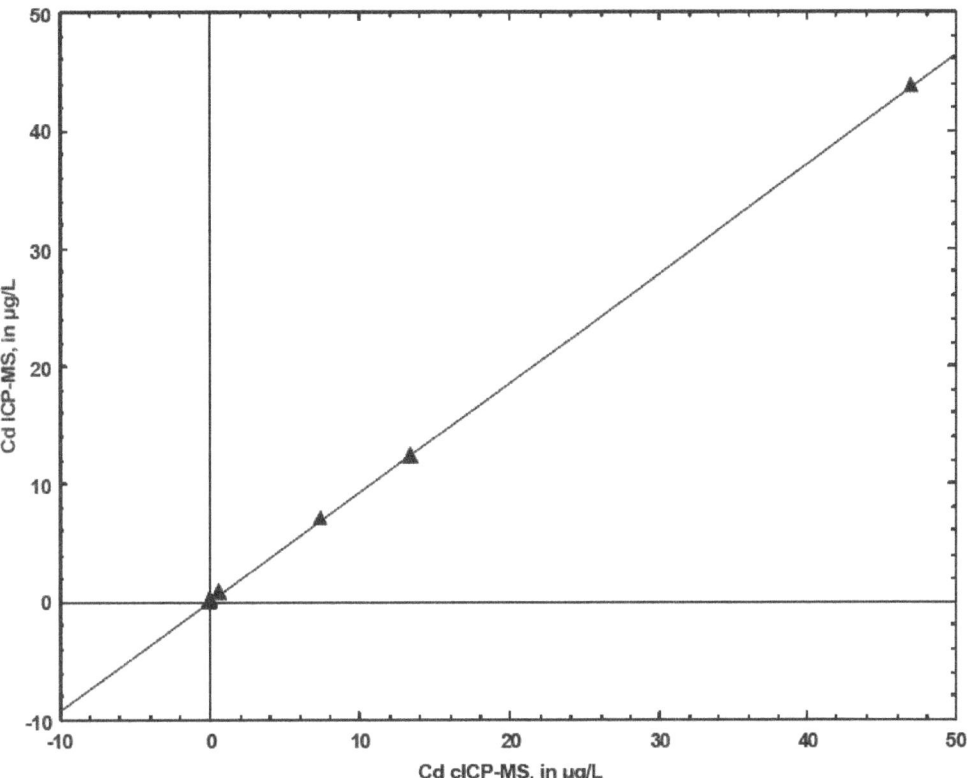

LINEAR REGRESSION EQUATION

Cd ICP-MS, in µg/L = .002 + .926 * Cd cICP-MS, in µg/L; R^2 = 1

Confidence Intervals
Cd ICP-MS, in µg/L vs. Cd cICP-MS, in µg/L

	Coefficient	95% Lower	95% Upper
Intercept	.002	-.020	.023
Cd cICP-MS, in µg/L	.926	.924	.928

Figure A15. Linear regression analysis of cadmium results from filtered water samples analyzed by inductively coupled plasma–mass spectrometry (ICP–MS) and collision/reaction cell inductively coupled plasma–mass spectrometry (cICP–MS). R^2 is the coefficient of determination. In the confidence intervals table, the Intercept and Cd cICP–MS coefficients are the y-intercept and slope, respectively. Results are in micrograms per liter (µg/L).

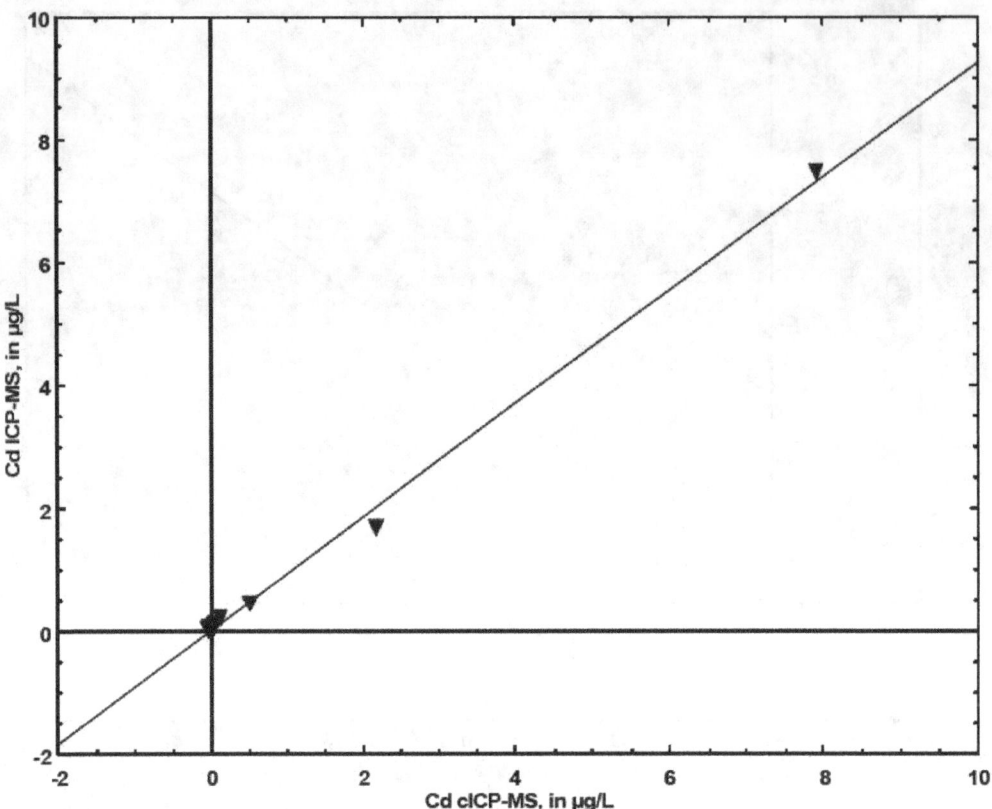

LINEAR REGRESSION EQUATION

Cd ICP-MS, in µg/L = -.002 + .925 * Cd cICP-MS, in µg/L; R^2 = .997

Confidence Intervals
Cd ICP-MS, in µg/L vs. Cd cICP-MS, in µg/L

	Coefficient	95% Lower	95% Upper
Intercept	-.002	-.036	.032
Cd cICP-MS, in µg/L	.925	.904	.946

Figure A16. Linear regression analysis of cadmium results from whole-water recoverable digestates analyzed by inductively coupled plasma–mass spectrometry (ICP–MS) and collision/reaction cell inductively coupled plasma–mass spectrometry (cICP–MS). R^2 is the coefficient of determination. In the confidence intervals table, the Intercept and Cd cICP–MS coefficients are the *y*-intercept and slope, respectively. Results are in micrograms per liter (µg/L).

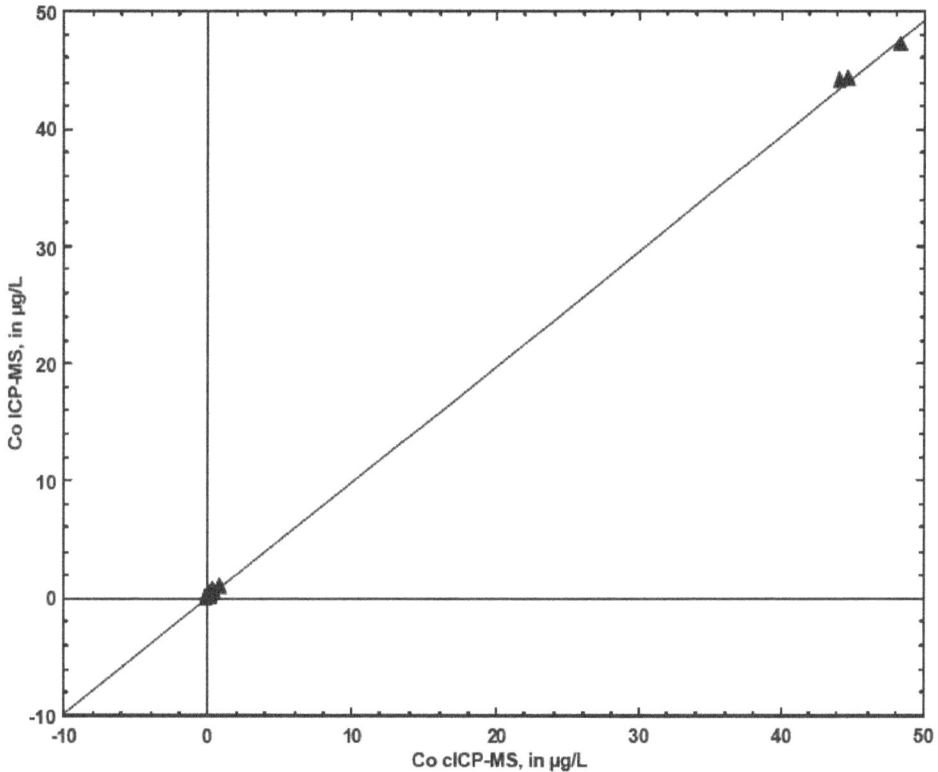

LINEAR REGRESSION EQUATION

Co ICP-MS, in µg/L = .018 + .986 * Co cICP-MS, in µg/L; R^2 = 1

Confidence Intervals
Co ICP-MS, in µg/L vs. Co cICP-MS, in µg/L

	Coefficient	95% Lower	95% Upper
Intercept	.018	-.049	.085
Co cICP-MS, in µg/L	.986	.981	.990

Figure A17. Linear regression analysis of cobalt results from filtered water samples analyzed by inductively coupled plasma–mass spectrometry (ICP–MS) and collision/reaction cell inductively coupled plasma–mass spectrometry (cICP–MS). R^2 is the coefficient of determination. In the confidence intervals table, the Intercept and Co cICP–MS coefficients are the y-intercept and slope, respectively. Results are in micrograms per liter (µg/L).

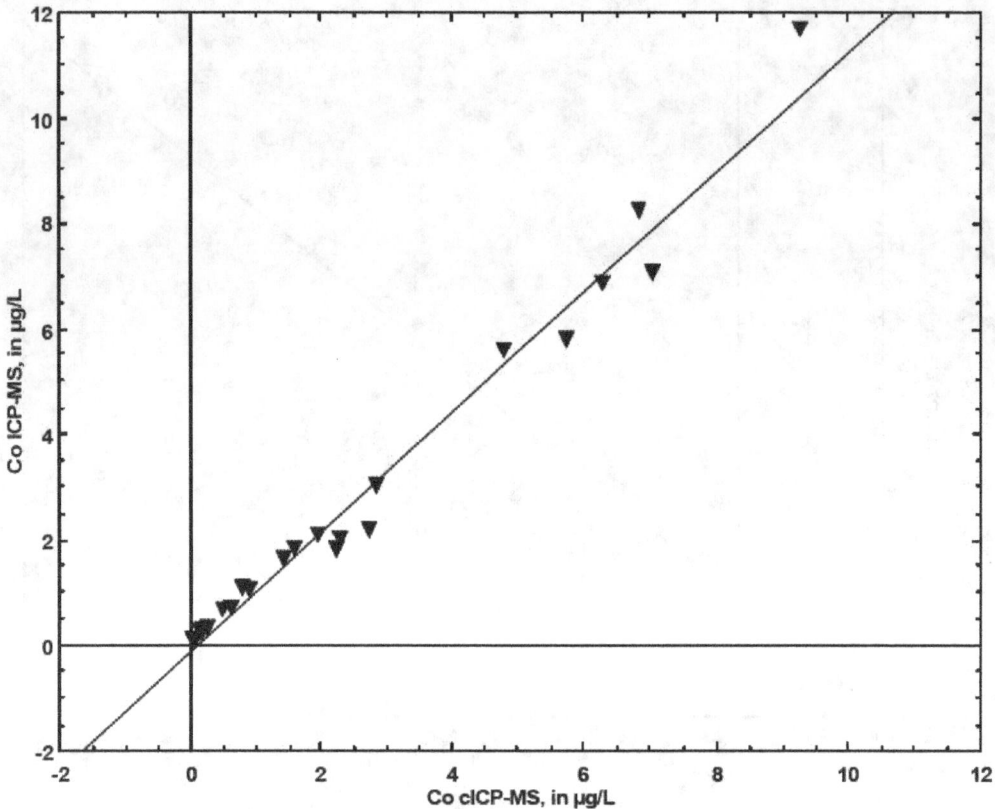

LINEAR REGRESSION EQUATION

Co ICP-MS, in µg/L = -.127 + 1.137 * Co cICP-MS, in µg/L; R^2 = .979

Confidence Intervals
Co ICP-MS, in µg/L vs. Co cICP-MS, in µg/L

	Coefficient	95% Lower	95% Upper
Intercept	-.127	-.382	.127
Co cICP-MS, in µg/L	1.137	1.065	1.208

Figure A18. Linear regression analysis of cobalt results from whole-water recoverable digestates analyzed by inductively coupled plasma–mass spectrometry (ICP–MS) and collision/reaction cell inductively coupled plasma–mass spectrometry (cICP–MS). R^2 is the coefficient of determination. In the confidence intervals table, the Intercept and Co cICP–MS coefficients are the y-intercept and slope, respectively. Results are in micrograms per liter (µg/L).

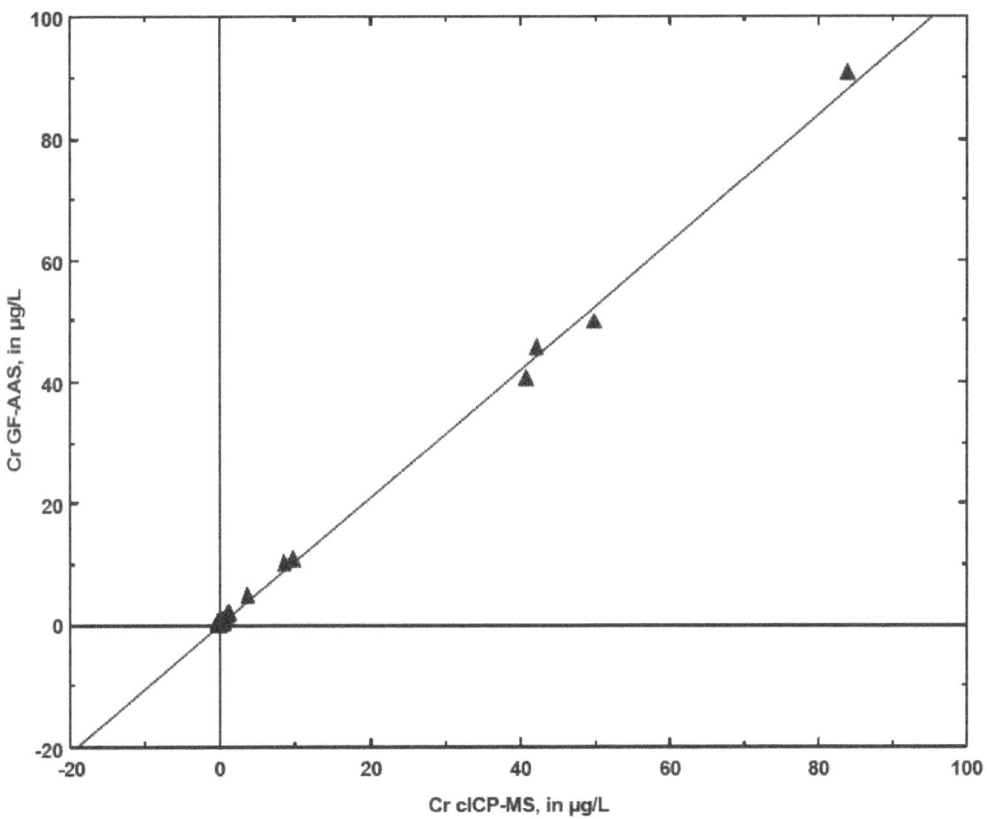

LINEAR REGRESSION EQUATION

Cr GF-AAS, in µg/L = -.003 + 1.05 * Cr cICP-MS, in µg/L; R^2 = .998

Confidence Intervals
Cr GF-AAS, in µg/L vs. Cr cICP-MS, in µg/L

	Coefficient	95% Lower	95% Upper
Intercept	-.003	-.489	.482
Cr cICP-MS, in µg/L	1.050	1.029	1.071

Figure A19. Linear regression analysis of chromium results from filtered water samples analyzed by graphite furnace–atomic absorption spectrometry (GF–AAS) and collision/reaction cell inductively coupled plasma–mass spectrometry (cICP–MS). R^2 is the coefficient of determination. In the confidence intervals table, the Intercept and Cr cICP–MS coefficients are the y-intercept and slope, respectively. Results are in micrograms per liter (µg/L).

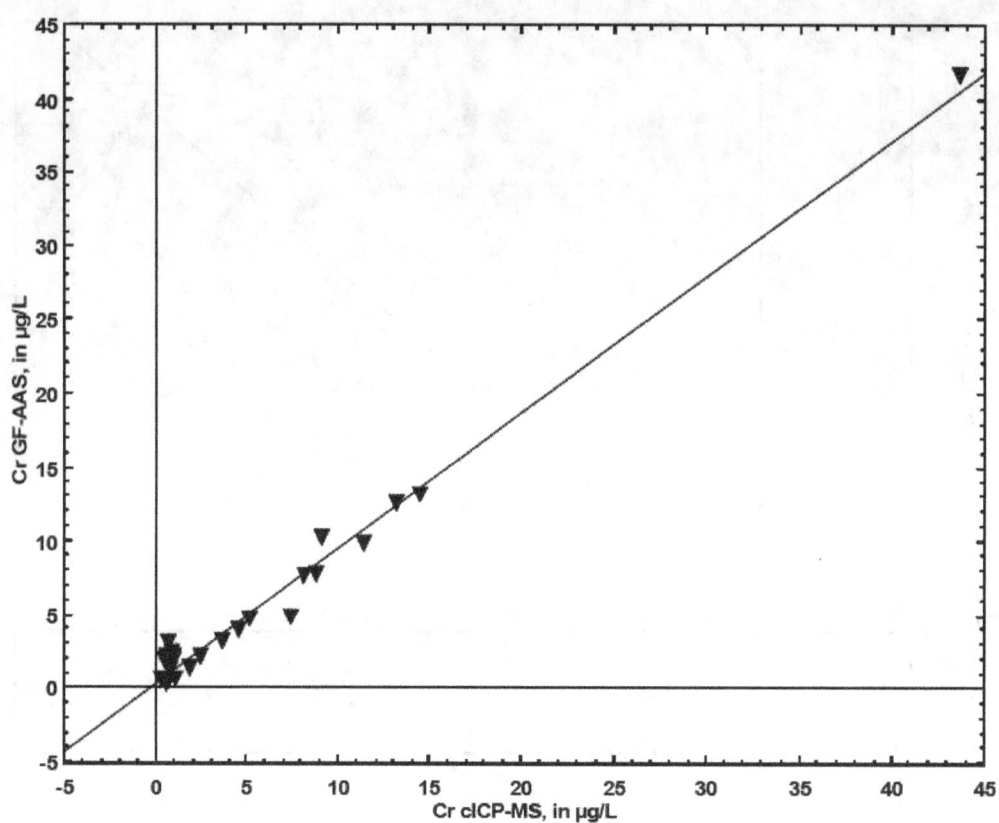

LINEAR REGRESSION EQUATION

Cr GF-AAS, in µg/L = .232 + .921 * Cr cICP-MS, in µg/L; R^2 = .986

Confidence Intervals
Cr GF-AAS, in µg/L vs. Cr cICP-MS, in µg/L

	Coefficient	95% Lower	95% Upper
Intercept	.232	-.274	.738
Cr cICP-MS, in µg/L	.921	.874	.969

Figure A20. Linear regression analysis of chromium results from whole-water recoverable digestates analyzed by graphite furnace–atomic absorption spectrometry (GF–AAS) and collision/reaction cell inductively coupled plasma–mass spectrometry (cICP–MS). R^2 is the coefficient of determination. In the confidence intervals table, the Intercept and Cr cICP–MS coefficients are the y-intercept and slope, respectively. Results are in micrograms per liter (µg/L).

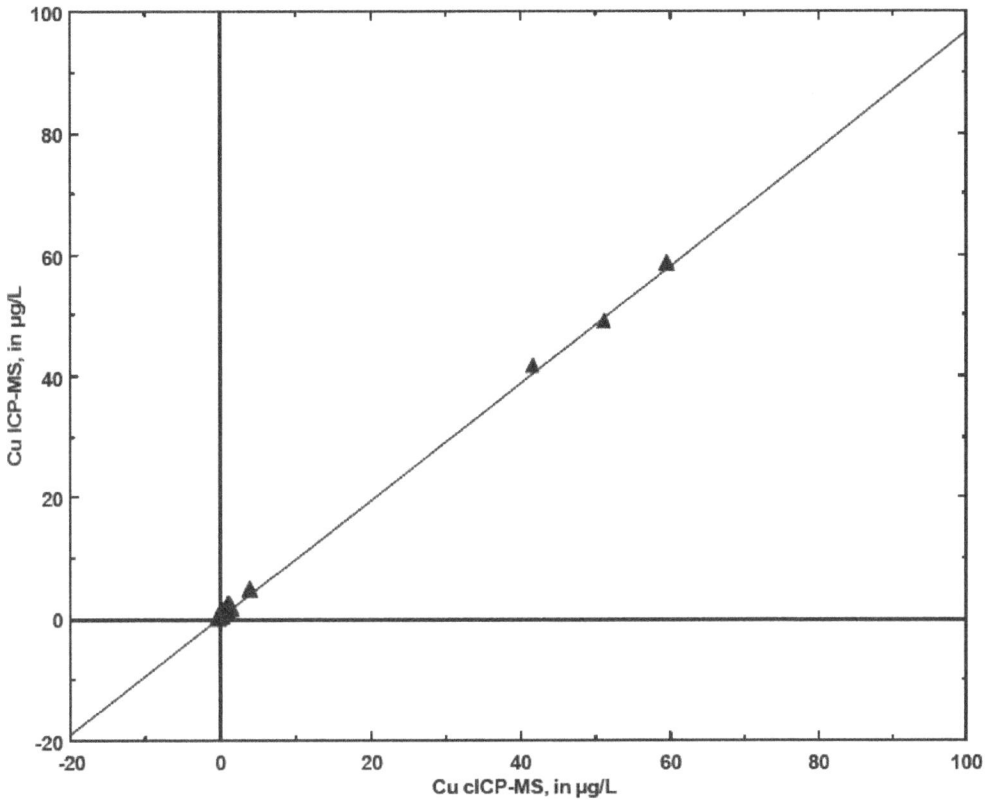

LINEAR REGRESSION EQUATION

Cu ICP-MS, in µg/L = .299 + .964 * Cu cICP-MS, in µg/L; R^2 = .999

Confidence Intervals
Cu ICP-MS, in µg/L vs. Cu cICP-MS, in µg/L

	Coefficient	95% Lower	95% Upper
Intercept	.299	.098	.499
Cu cICP-MS, in µg/L	.964	.953	.975

Figure A21. Linear regression analysis of copper results from filtered water samples analyzed by inductively coupled plasma–mass spectrometry (ICP–MS) and collision/reaction cell inductively coupled plasma–mass spectrometry (cICP–MS). R^2 is the coefficient of determination. In the confidence intervals table, the Intercept and Cu cICP–MS coefficients are the *y*-intercept and slope, respectively. Results are in micrograms per liter (µg/L).

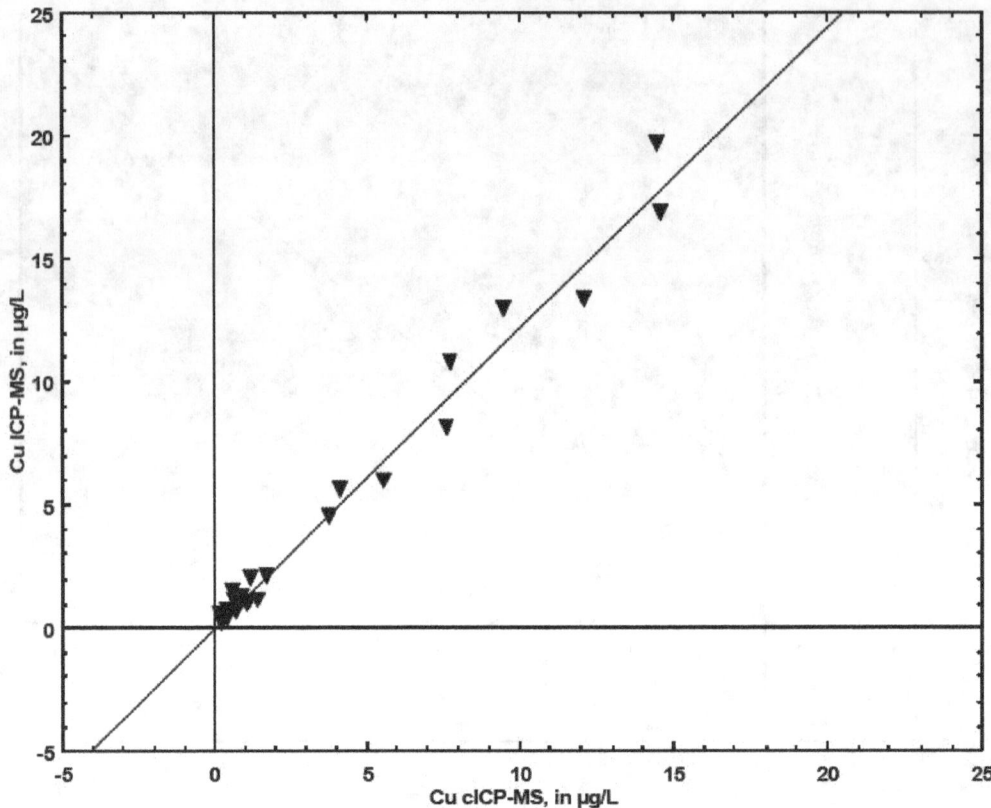

LINEAR REGRESSION EQUATION

Cu ICP-MS, in µg/L = -.034 + 1.22 * Cu cICP-MS, in µg/L; R^2 = .981

Confidence Intervals
Cu ICP-MS, in µg/L vs. Cu cICP-MS, in µg/L

	Coefficient	95% Lower	95% Upper
Intercept	-.034	-.463	.396
Cu cICP-MS, in µg/L	1.220	1.147	1.294

Figure A22. Linear regression analysis of copper results from whole-water recoverable digestates analyzed by inductively coupled plasma–mass spectrometry (ICP–MS) and collision/reaction cell inductively coupled plasma–mass spectrometry (cICP–MS). R^2 is the coefficient of determination. In the confidence intervals table, the Intercept and Cu cICP–MS coefficients are the y-intercept and slope, respectively. Results are in micrograms per liter (µg/L).

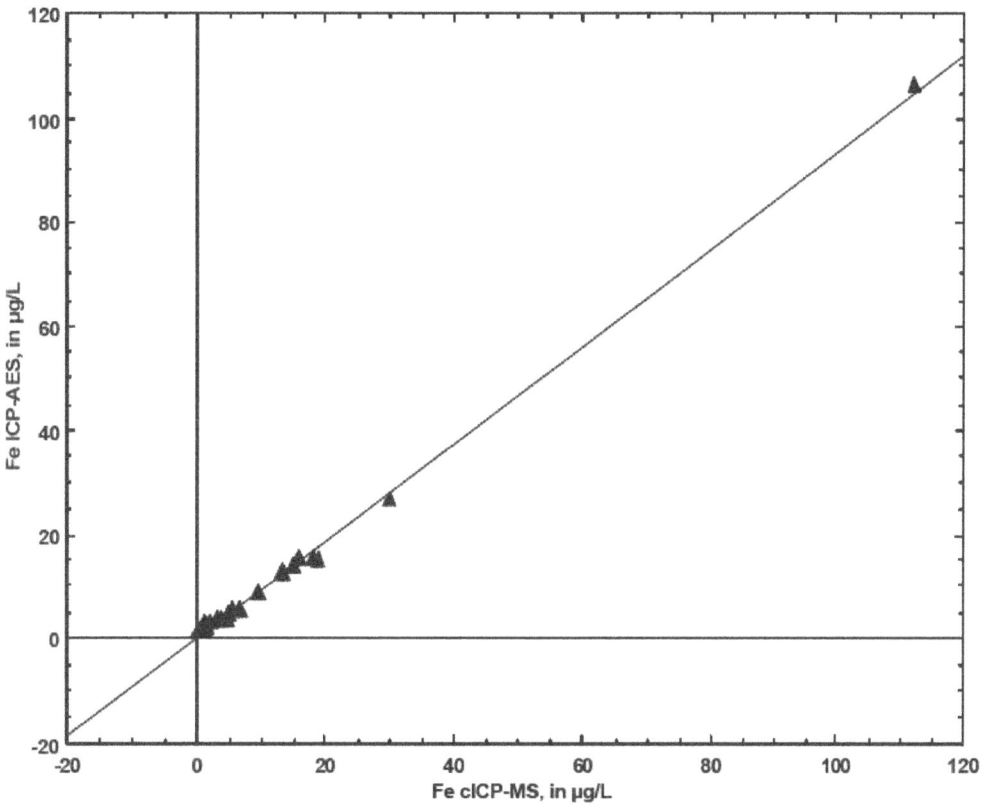

LINEAR REGRESSION EQUATION

Fe ICP-AES, in µg/L = -.035 + .934 * Fe cICP-MS, in µg/L; R^2 = .997

Confidence Intervals
Fe ICP-AES, in µg/L vs. Fe cICP-MS, in µg/L

	Coefficient	95% Lower	95% Upper
Intercept	-.035	-.549	.480
Fe cICP-MS, in µg/L	.934	.913	.955

Figure A23. Linear regression analysis of iron results from filtered water samples analyzed by inductively coupled plasma–atomic emission spectrometry (ICP–AES) and collision/reaction cell inductively coupled plasma–mass spectrometry (cICP–MS). R^2 is the coefficient of determination. In the confidence intervals table, the Intercept and Fe cICP–MS coefficients are the *y*-intercept and slope, respectively. Results are in micrograms per liter (µg/L).

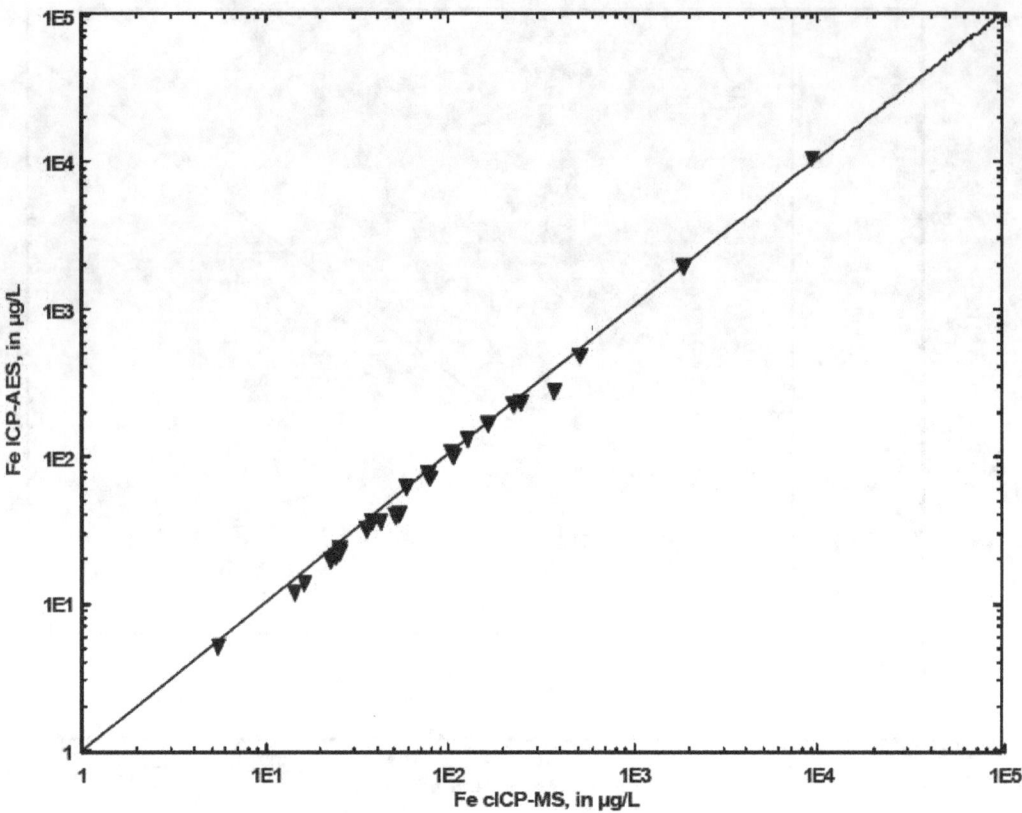

LINEAR REGRESSION EQUATION

Fe ICP-AES, in µg/L = -26.485 + 1.078 * Fe cICP-MS, in µg/L; $R^2 = 1$

Confidence Intervals
Fe ICP-AES, in µg/L vs. Fe cICP-MS, in µg/L

	Coefficient	95% Lower	95% Upper
Intercept	-26.485	-44.006	-8.963
Fe cICP-MS, in µg/L	1.078	1.069	1.087

Figure A24. Linear regression analysis of iron results from whole-water recoverable digestates analyzed by inductively coupled plasma–atomic emission spectrometry (ICP–AES) and collision/reaction cell inductively coupled plasma–mass spectrometry (cICP–MS). R^2 is the coefficient of determination. In the confidence intervals table, the Intercept and Fe cICP–MS coefficients are the y-intercept and slope, respectively. Results are in micrograms per liter (µg/L).

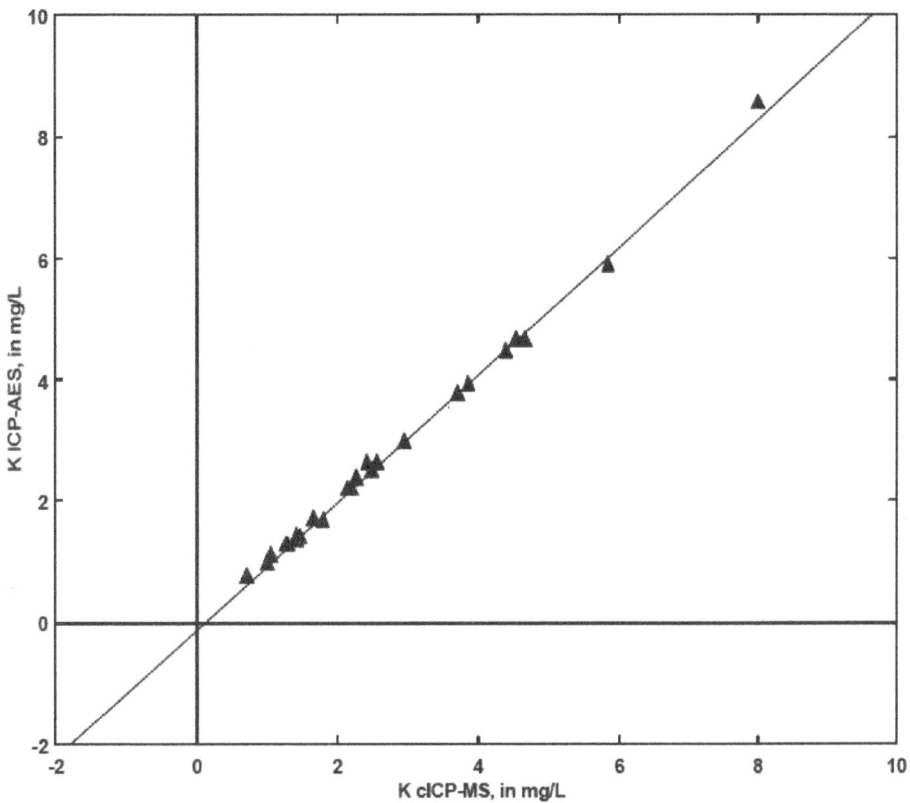

LINEAR REGRESSION EQUATION

K ICP-AES, in mg/L = -.112 + 1.048 * K cICP-MS, in mg/L; R^2 = .998

Confidence Intervals
K ICP-AES, in mg/L vs. K cICP-MS, in mg/L

	Coefficient	95% Lower	95% Upper
Intercept	-.112	-.183	-.040
K cICP-MS, in mg/L	1.048	1.025	1.070

Figure A25. Linear regression analysis of potassium results from filtered water samples analyzed by inductively coupled plasma–atomic emission spectrometry (ICP–AES) and collision/reaction cell inductively coupled plasma–mass spectrometry (cICP–MS). R^2 is the coefficient of determination. In the confidence intervals table, the Intercept and K cICP–MS coefficients are the *y*-intercept and slope, respectively. Results are in milligrams per liter (mg/L).

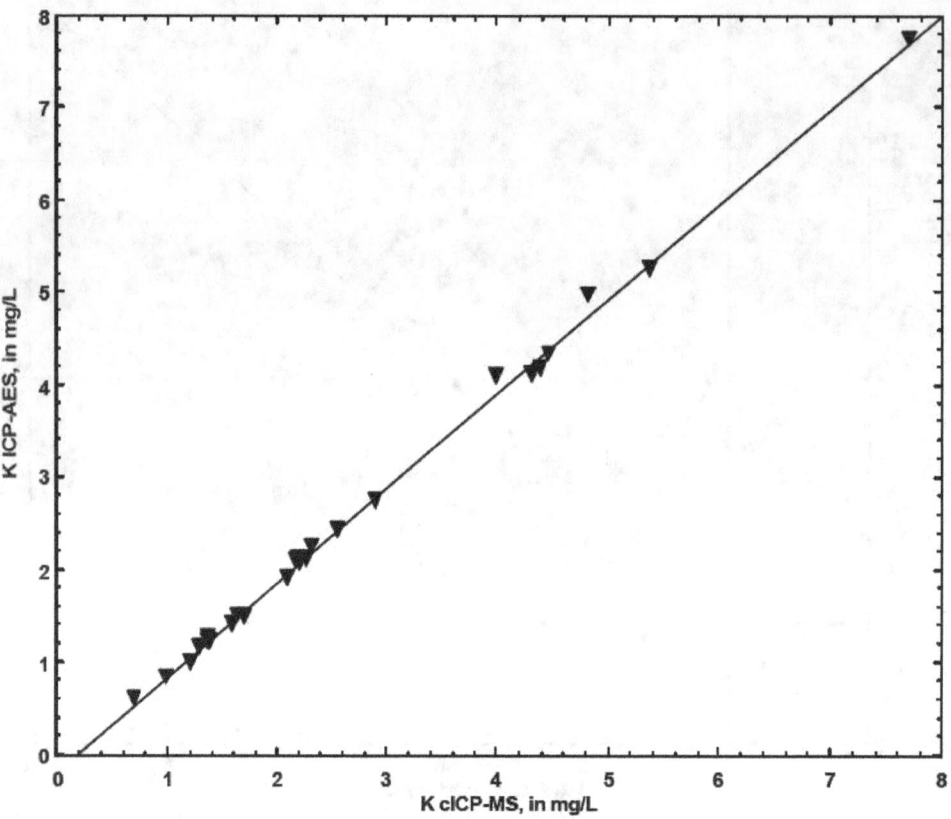

LINEAR REGRESSION EQUATION

K ICP-AES, in mg/L = -.21 + 1.026 * K cICP-MS, in mg/L; R^2 = .998

Confidence Intervals
K ICP-AES, in mg/L vs. K cICP-MS, in mg/L

	Coefficient	95% Lower	95% Upper
Intercept	-.210	-.276	-.145
K cICP-MS, in mg/L	1.026	1.005	1.047

Figure A26. Linear regression analysis of potassium results from whole-water recoverable digestates analyzed by inductively coupled plasma–atomic emission spectrometry (ICP–AES) and collision/reaction cell inductively coupled plasma–mass spectrometry (cICP–MS). R^2 is the coefficient of determination. In the confidence intervals table, the Intercept and K cICP–MS coefficients are the y-intercept and slope, respectively. Results are in milligrams per liter (mg/L).

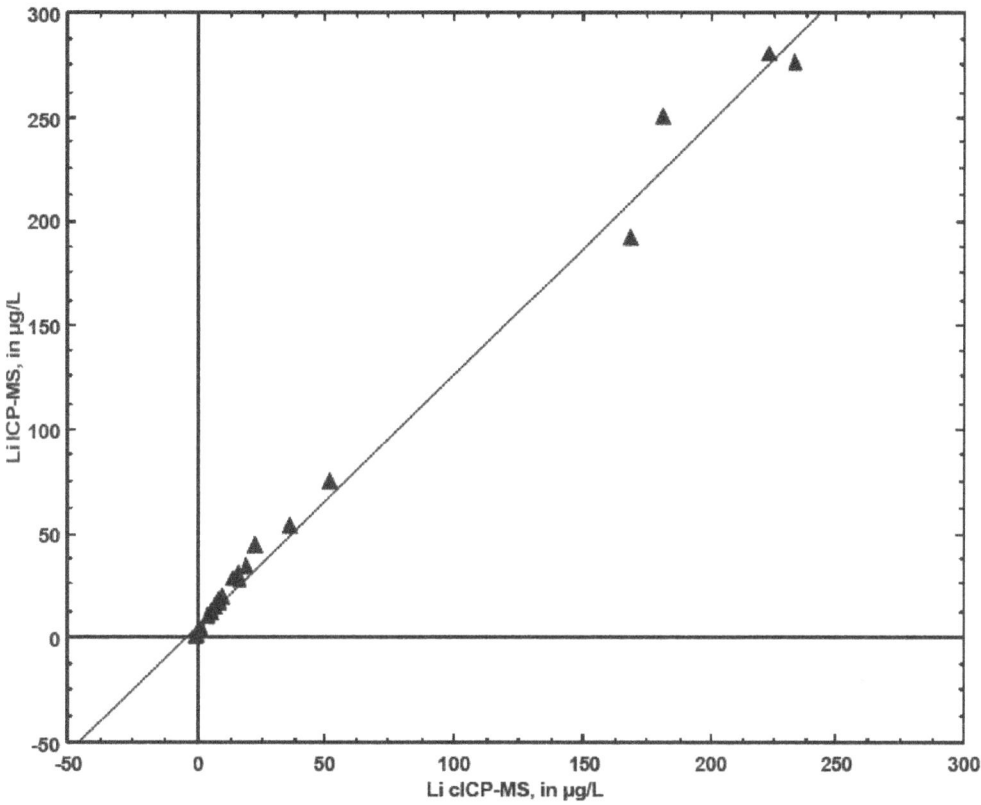

LINEAR REGRESSION EQUATION

Li ICP-MS, in µg/L = 5.659 + 1.208 * Li cICP-MS, in µg/L; R^2 = .992

Confidence Intervals
Li ICP-MS, in µg/L vs. Li cICP-MS, in µg/L

	Coefficient	95% Lower	95% Upper
Intercept	5.659	1.708	9.610
Li cICP-MS, in µg/L	1.208	1.160	1.255

Figure A27. Linear regression analysis of lithium results from filtered water samples analyzed by inductively coupled plasma–mass spectrometry (ICP–MS) and collision/reaction cell inductively coupled plasma–mass spectrometry (cICP–MS). R^2 is the coefficient of determination. In the confidence intervals table, the Intercept and Li cICP–MS coefficients are the *y*-intercept and slope, respectively. Results are in micrograms per liter (µg/L).

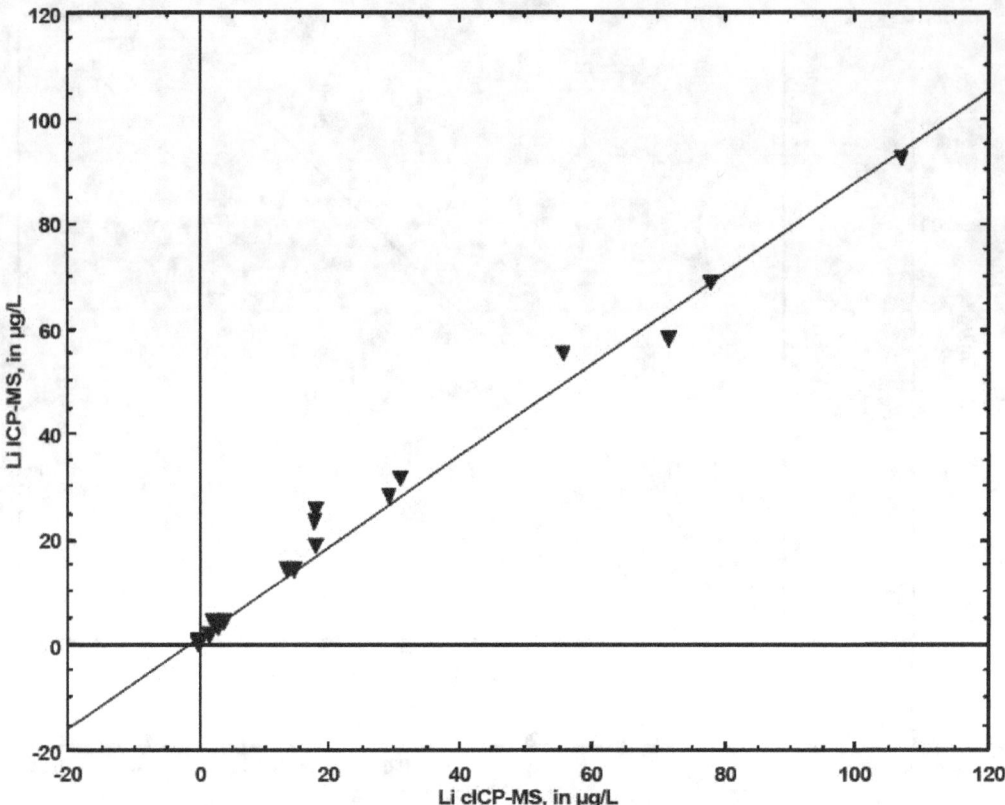

LINEAR REGRESSION EQUATION

Li ICP-MS, in µg/L = 1.472 + .863 * Li cICP-MS, in µg/L; R^2 = .987

Confidence Intervals
Li ICP-MS, in µg/L vs. Li cICP-MS, in µg/L

	Coefficient	95% Lower	95% Upper
Intercept	1.472	.018	2.926
Li cICP-MS, in µg/L	.863	.821	.906

Figure A28. Linear regression analysis of lithium results from whole-water recoverable digestates analyzed by inductively coupled plasma–mass spectrometry (ICP–MS) and collision/reaction cell inductively coupled plasma–mass spectrometry (cICP–MS). R^2 is the coefficient of determination. In the confidence intervals table, the Intercept and Li cICP–MS coefficients are the y-intercept and slope, respectively. Results are in micrograms per liter (µg/L).

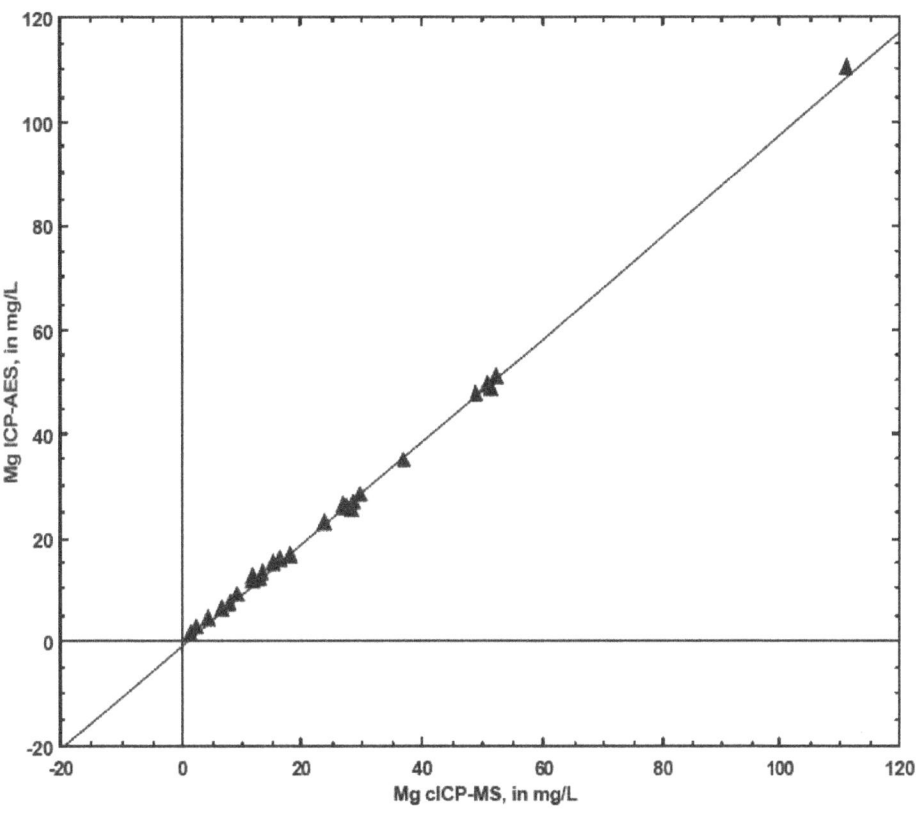

LINEAR REGRESSION EQUATION

Mg ICP-AES, in mg/L = -.779 + .982 * Mg cICP-MS, in mg/L; R^2 = .999

Confidence Intervals
Mg ICP-AES, in mg/L vs. Mg cICP-MS, in mg/L

	Coefficient	95% Lower	95% Upper
Intercept	-.779	-1.299	-.260
Mg cICP-MS, in mg/L	.982	.967	.997

Figure A29. Linear regression analysis of magnesium results from filtered water samples analyzed by inductively coupled plasma–atomic emission spectrometry (ICP–AES) and collision/reaction cell inductively coupled plasma–mass spectrometry (cICP–MS). R^2 is the coefficient of determination. In the confidence intervals table, the Intercept and Mg cICP–MS coefficients are the y-intercept and slope, respectively. Results are in milligrams per liter (mg/L).

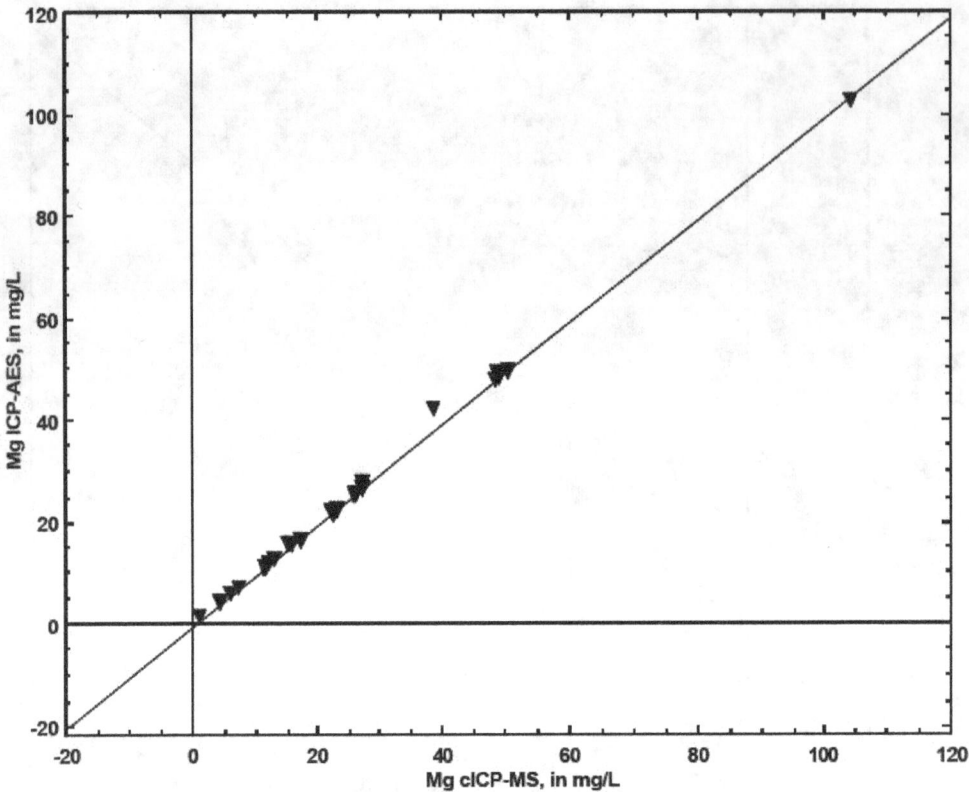

LINEAR REGRESSION EQUATION

Mg ICP-AES, in mg/L = -.594 + .993 * Mg cICP-MS, in mg/L; R^2 = .998

Confidence Intervals
Mg ICP-AES, in mg/L vs. Mg cICP-MS, in mg/L

	Coefficient	95% Lower	95% Upper
Intercept	-.594	-1.206	.019
Mg cICP-MS, in mg/L	.993	.975	1.011

Figure A30. Linear regression analysis of magnesium results from whole-water recoverable digestates analyzed by inductively coupled plasma–atomic emission spectrometry (ICP–AES) and collision/ reaction cell inductively coupled plasma–mass spectrometry (cICP–MS). R^2 is the coefficient of determination. In the confidence intervals table, the Intercept and Mg cICP–MS coefficients are the y-intercept and slope, respectively. Results are in milligrams per liter (mg/L).

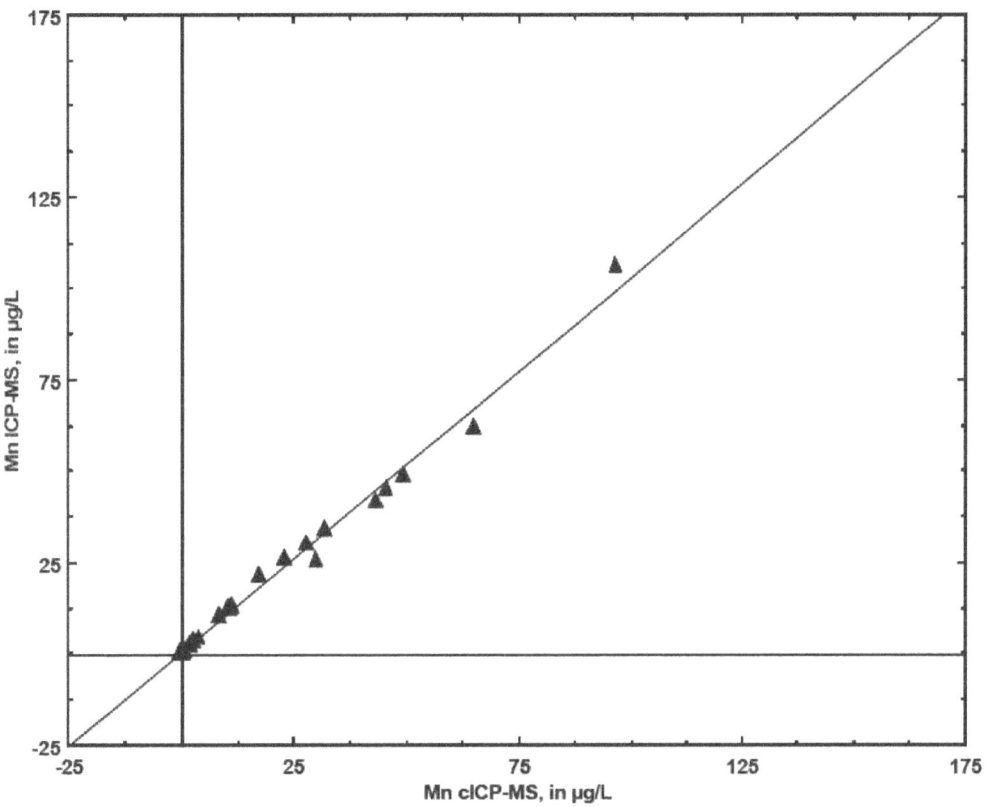

LINEAR REGRESSION EQUATION

Mn ICP-MS, in µg/L = .082 + 1.026 * Mn cICP-MS, in µg/L; R^2 = .991

Confidence Intervals
Mn ICP-MS, in µg/L vs. Mn cICP-MS, in µg/L

	Coefficient	95% Lower	95% Upper
Intercept	.082	-1.256	1.420
Mn cICP-MS, in µg/L	1.026	.983	1.069

Figure A31. Linear regression analysis of manganese results from filtered water samples analyzed by inductively coupled plasma–mass spectrometry (ICP–MS) and collision/reaction cell inductively coupled plasma–mass spectrometry (cICP–MS). R^2 is the coefficient of determination. In the confidence intervals table, the Intercept and Mn cICP–MS coefficients are the *y*-intercept and slope, respectively. Results are in micrograms per liter (µg/L).

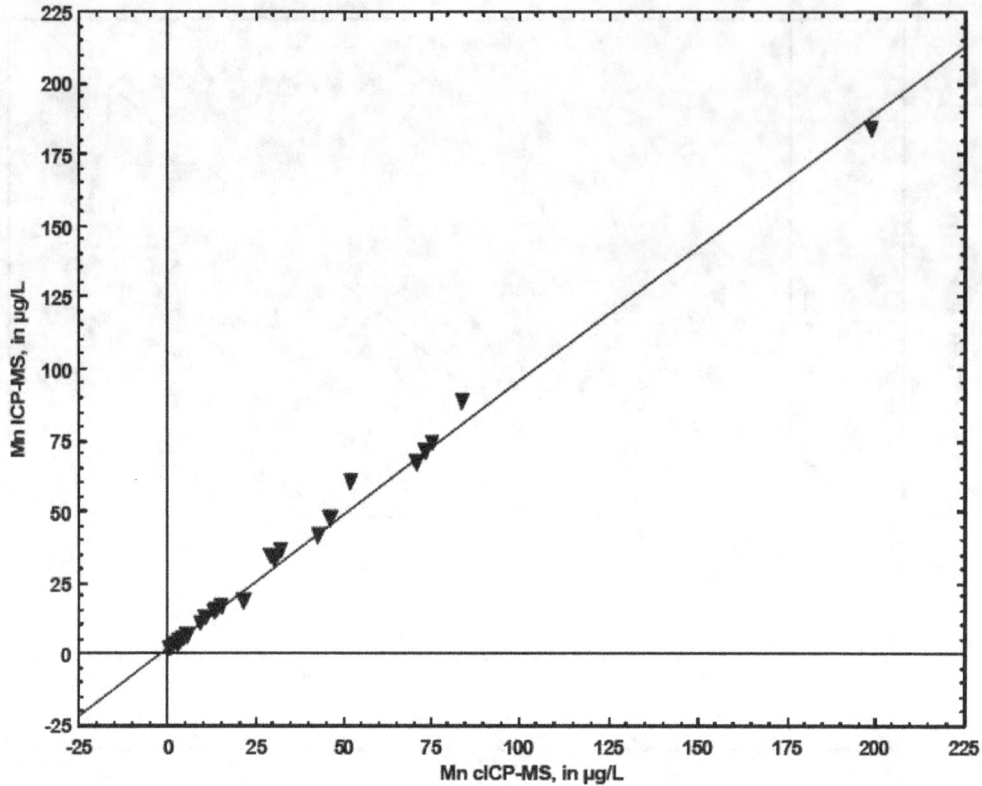

LINEAR REGRESSION EQUATION

Mn ICP-MS, in µg/L = 1.682 + .938 * Mn cICP-MS, in µg/L; R^2 = .994

Confidence Intervals
Mn ICP-MS, in µg/L vs. Mn cICP-MS, in µg/L

	Coefficient	95% Lower	95% Upper
Intercept	1.682	-.026	3.390
Mn cICP-MS, in µg/L	.938	.907	.969

Figure A32. Linear regression analysis of manganese results from whole-water recoverable digestates analyzed by inductively coupled plasma–mass spectrometry (ICP–MS) and collision/reaction cell inductively coupled plasma–mass spectrometry (cICP–MS). R^2 is the coefficient of determination. In the confidence intervals table, the Intercept and Mn cICP–MS coefficients are the *y*-intercept and slope, respectively. Results are in micrograms per liter (µg/L).

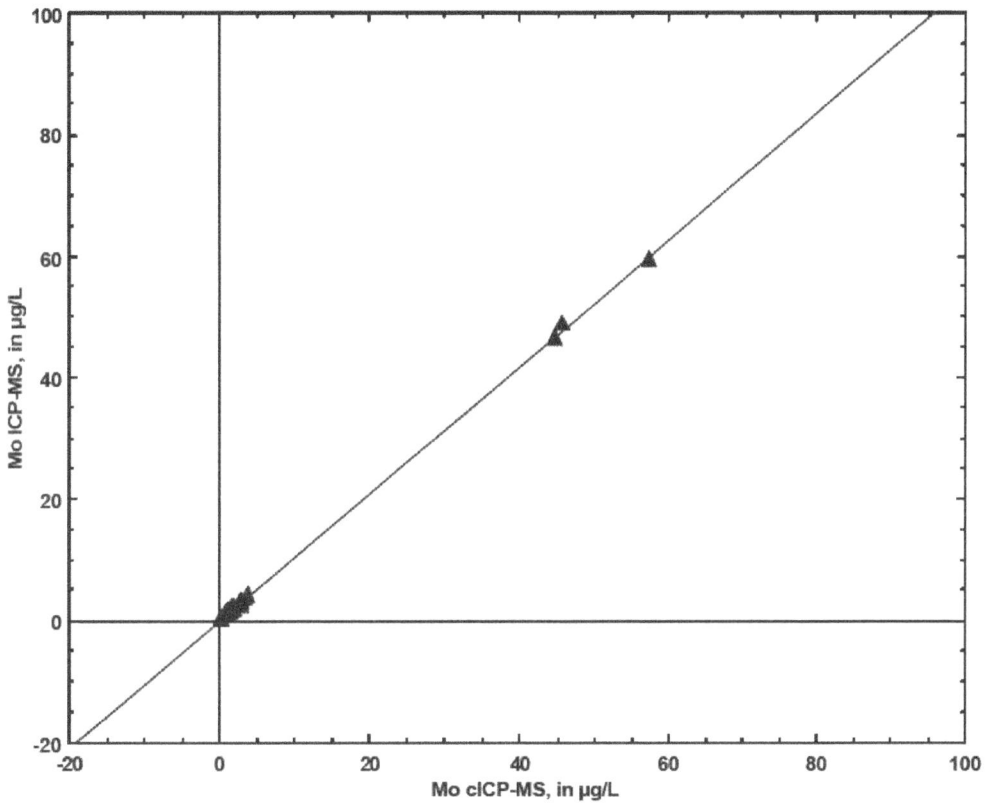

LINEAR REGRESSION EQUATION

Mo ICP-MS, in µg/L = -.112 + 1.045 * Mo cICP-MS, in µg/L; R^2 = 1

Confidence Intervals
Mo ICP-MS, in µg/L vs. Mo cICP-MS, in µg/L

	Coefficient	95% Lower	95% Upper
Intercept	-.112	-.239	.016
Mo cICP-MS, in µg/L	1.045	1.038	1.053

Figure A33. Linear regression analysis of molybdenum results from filtered water samples analyzed by inductively coupled plasma–mass spectrometry (ICP–MS) and collision/reaction cell inductively coupled plasma–mass spectrometry (cICP–MS). R^2 is the coefficient of determination. In the confidence intervals table, the Intercept and Mo cICP–MS coefficients are the y-intercept and slope, respectively. Results are in micrograms per liter (µg/L).

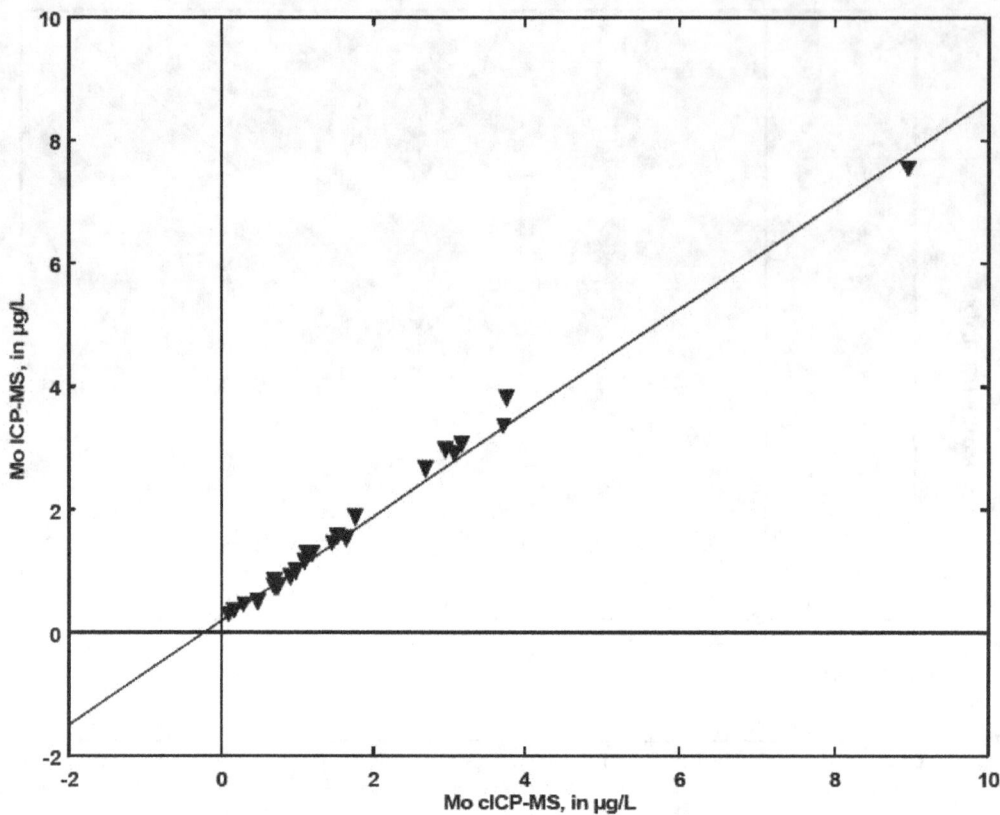

LINEAR REGRESSION EQUATION

Mo ICP-MS, in µg/L = .202 + .847 * Mo cICP-MS, in µg/L; R^2 = .993

Confidence Intervals
Mo ICP-MS, in µg/L vs. Mo cICP-MS, in µg/L

	Coefficient	95% Lower	95% Upper
Intercept	.202	.119	.286
Mo cICP-MS, in µg/L	.847	.815	.879

Figure A34. Linear regression analysis of molybdenum results from whole-water recoverable digestates analyzed by inductively coupled plasma–mass spectrometry (ICP–MS) and collision/reaction cell inductively coupled plasma–mass spectrometry (cICP–MS). R^2 is the coefficient of determination. In the confidence intervals table, the Intercept and Mo cICP–MS coefficients are the y-intercept and slope, respectively. Results are in micrograms per liter (µg/L).

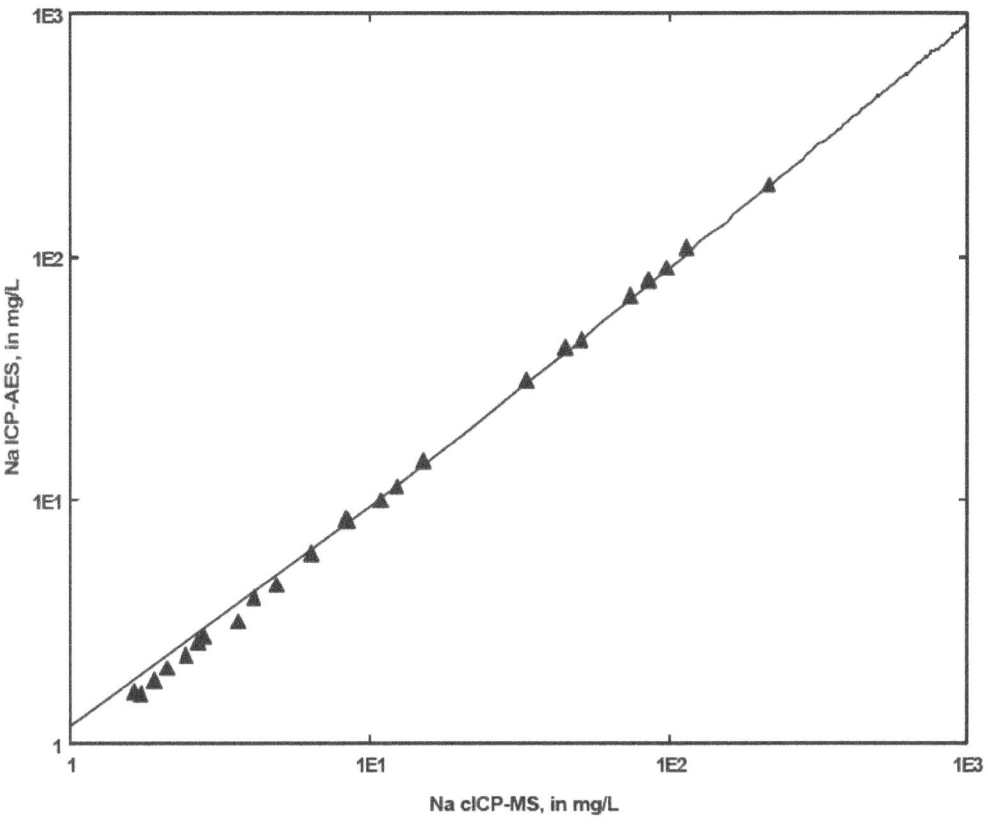

LINEAR REGRESSION EQUATION

Na ICP-AES, in mg/L = .273 + .906 * Na cICP-MS, in mg/L; R^2 = 1

Confidence Intervals
Na ICP-AES, in mg/L vs. Na cICP-MS, in mg/L

	Coefficient	95% Lower	95% Upper
Intercept	.273	-.263	.810
Na cICP-MS, in mg/L	.906	.898	.915

Figure A35. Linear regression analysis of sodium results from filtered water samples analyzed by inductively coupled plasma–atomic emission spectrometry (ICP–AES) and collision/reaction cell inductively coupled plasma–mass spectrometry (cICP–MS). R^2 is the coefficient of determination. In the confidence intervals table, the Intercept and Na cICP–MS coefficients are the y-intercept and slope, respectively. Results are in milligrams per liter (mg/L).

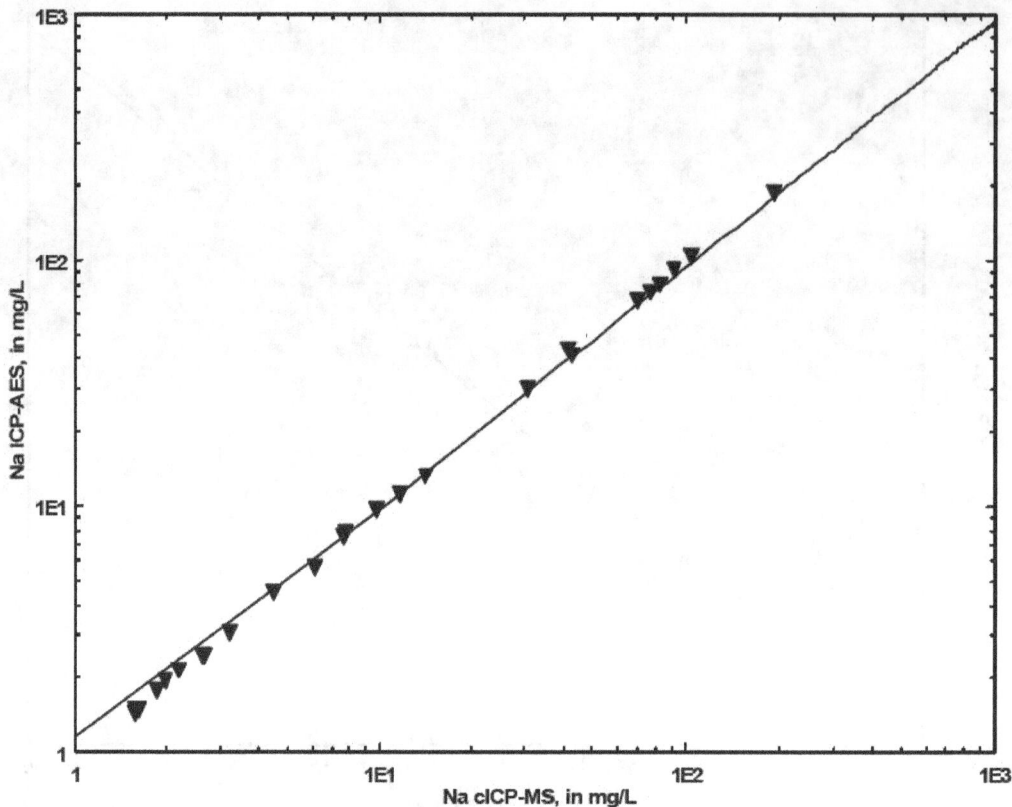

LINEAR REGRESSION EQUATION

Na ICP-AES, in mg/L = .213 + .942 * Na cICP-MS, in mg/L; R^2 = .999

Confidence Intervals
Na ICP-AES, in mg/L vs. Na cICP-MS, in mg/L

	Coefficient	95% Lower	95% Upper
Intercept	.213	-.382	.807
Na cICP-MS, in mg/L	.942	.932	.953

Figure A36. Linear regression analysis of sodium results from whole-water recoverable digestates analyzed by inductively coupled plasma–atomic emission spectrometry (ICP–AES) and collision/reaction cell inductively coupled plasma–mass spectrometry (cICP–MS). R^2 is the coefficient of determination. In the confidence intervals table, the Intercept and Na cICP–MS coefficients are the y-intercept and slope, respectively. Results are in milligrams per liter (mg/L).

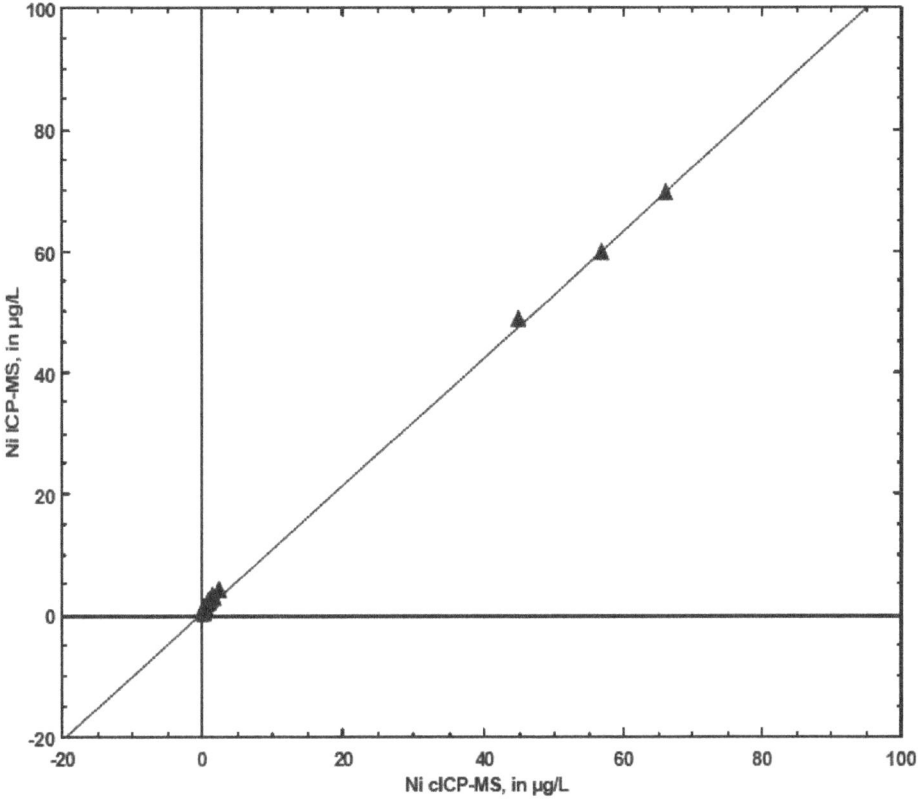

LINEAR REGRESSION EQUATION

Ni ICP-MS, in μg/L = .494 + 1.045 * Ni cICP-MS, in μg/L; R^2 = .999

Confidence Intervals
Ni ICP-MS, in μg/L vs. Ni cICP-MS, in μg/L

	Coefficient	95% Lower	95% Upper
Intercept	.494	.285	.704
Ni cICP-MS, in μg/L	1.045	1.035	1.056

Figure A37. Linear regression analysis of nickel results from filtered water samples analyzed by inductively coupled plasma–mass spectrometry (ICP–MS) and collision/reaction cell inductively coupled plasma–mass spectrometry (cICP–MS). R^2 is the coefficient of determination. In the confidence intervals table, the Intercept and Ni cICP–MS coefficients are the y-intercept and slope, respectively. Results are in micrograms per liter (μg/L).

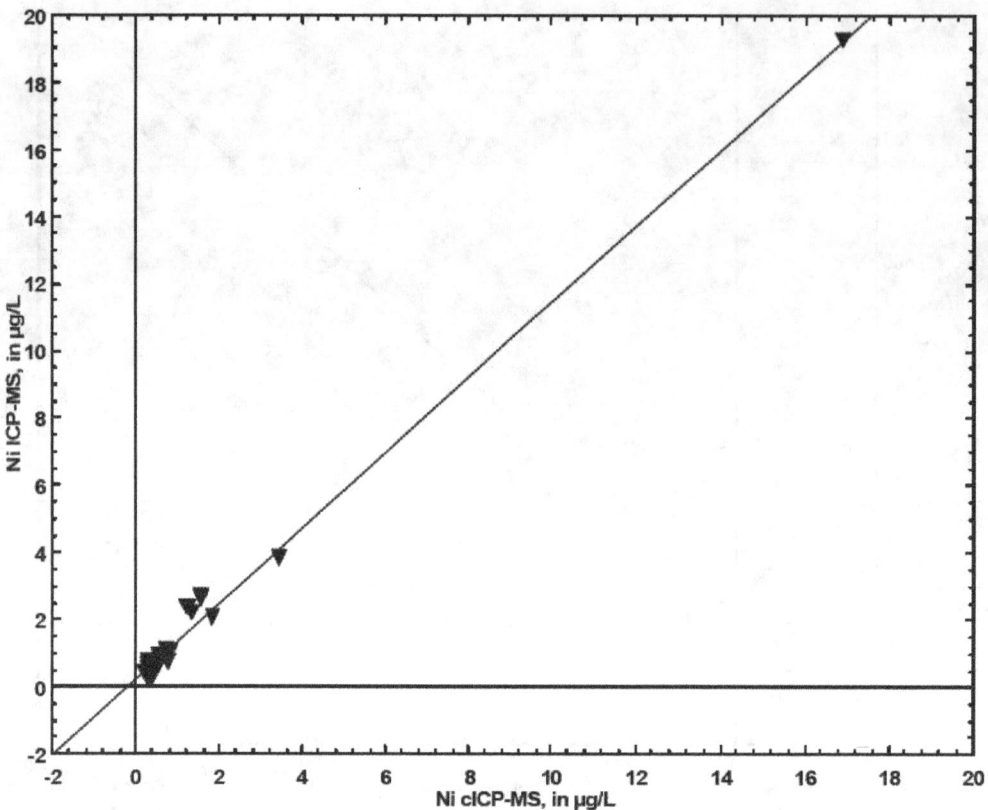

LINEAR REGRESSION EQUATION

Ni ICP-MS, in µg/L = .188 + 1.128 * Ni cICP-MS, in µg/L; R^2 = .99

Confidence Intervals
Ni ICP-MS, in µg/L vs. Ni cICP-MS, in µg/L

	Coefficient	95% Lower	95% Upper
Intercept	.188	.016	.360
Ni cICP-MS, in µg/L	1.128	1.080	1.176

Figure A38. Linear regression analysis of nickel results from whole-water recoverable digestates analyzed by inductively coupled plasma–mass spectrometry (ICP–MS) and collision/reaction cell inductively coupled plasma–mass spectrometry (cICP–MS). R^2 is the coefficient of determination. In the confidence intervals table, the Intercept and Ni cICP–MS coefficients are the y-intercept and slope, respectively. Results are in micrograms per liter (µg/L).

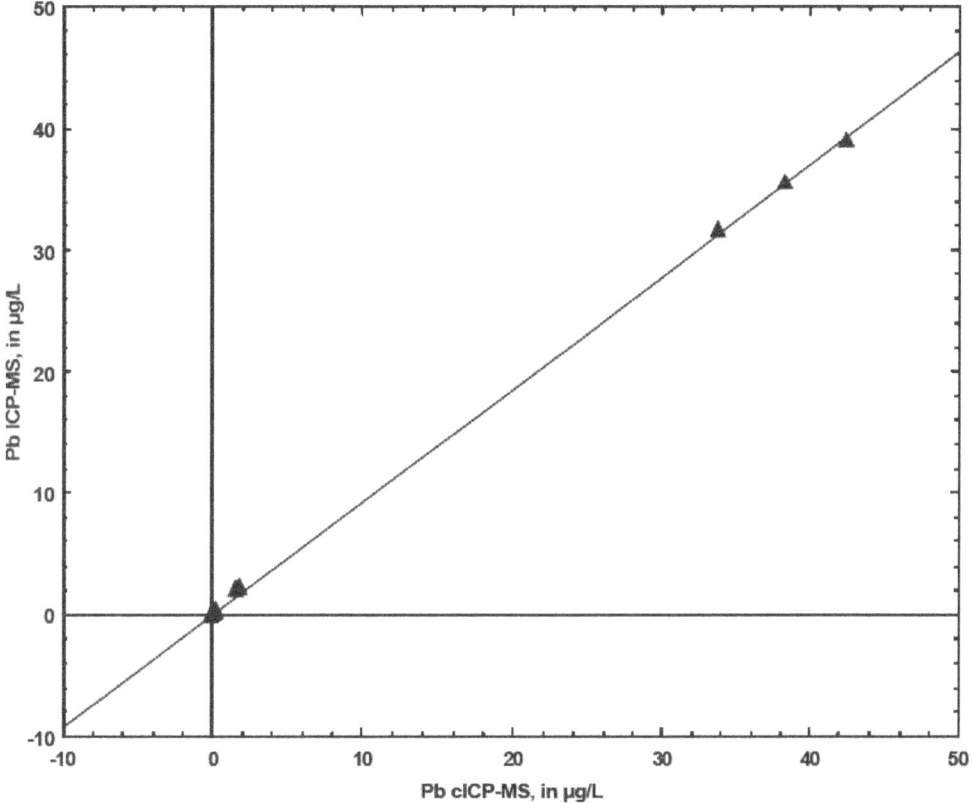

LINEAR REGRESSION EQUATION

Pb ICP-MS, in µg/L = .024 + .922 * Pb cICP-MS, in µg/L; R^2 = 1

Confidence Intervals
Pb ICP-MS, in µg/L vs. Pb cICP-MS, in µg/L

	Coefficient	95% Lower	95% Upper
Intercept	.024	-.047	.095
Pb cICP-MS, in µg/L	.922	.917	.928

Figure A39. Linear regression analysis of lead results from filtered water samples analyzed by inductively coupled plasma–mass spectrometry (ICP–MS) and collision/reaction cell inductively coupled plasma–mass spectrometry (cICP–MS). R^2 is the coefficient of determination. In the confidence intervals table, the Intercept and Pb cICP–MS coefficients are the y-intercept and slope, respectively. Results are in micrograms per liter (µg/L).

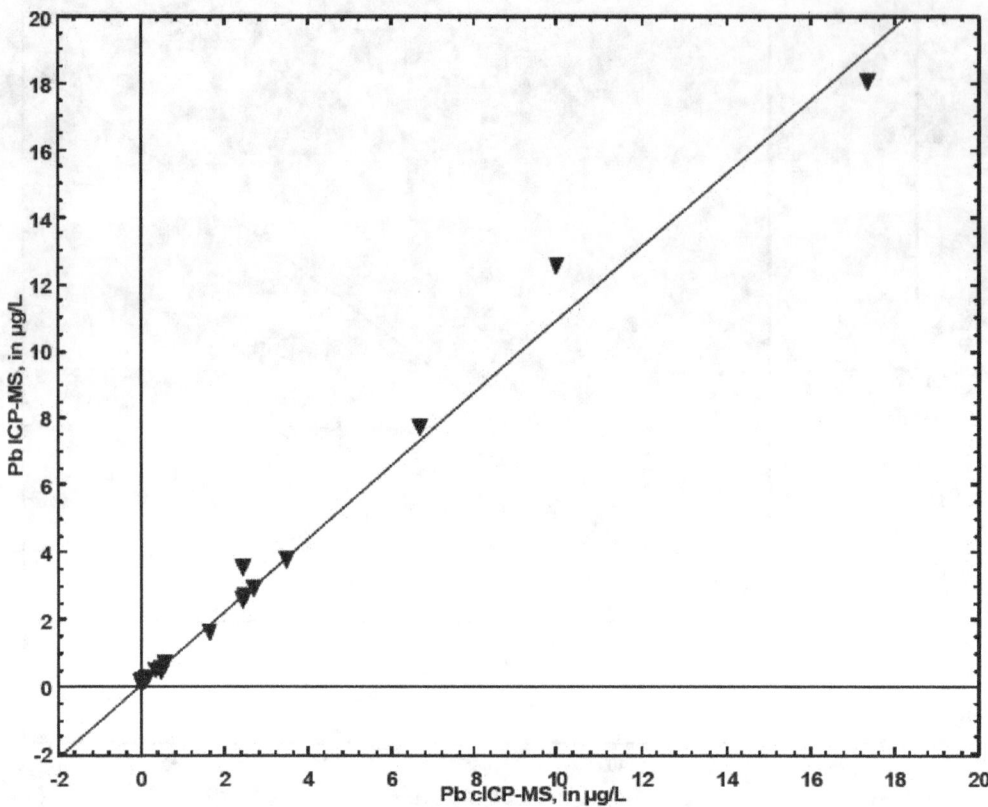

LINEAR REGRESSION EQUATION

Pb ICP-MS, in µg/L = .039 + 1.089 * Pb cICP-MS, in µg/L; R^2 = .99

Confidence Intervals
Pb ICP-MS, in µg/L vs. Pb cICP-MS, in µg/L

	Coefficient	95% Lower	95% Upper
Intercept	.039	-.167	.244
Pb cICP-MS, in µg/L	1.089	1.043	1.135

Figure A40. Linear regression analysis of lead results from whole-water recoverable digestates analyzed by inductively coupled plasma–mass spectrometry (ICP–MS) and collision/reaction cell inductively coupled plasma–mass spectrometry (cICP–MS). R^2 is the coefficient of determination. In the confidence intervals table, the Intercept and Pb cICP–MS coefficients are the *y*-intercept and slope, respectively. Results are in micrograms per liter (µg/L).

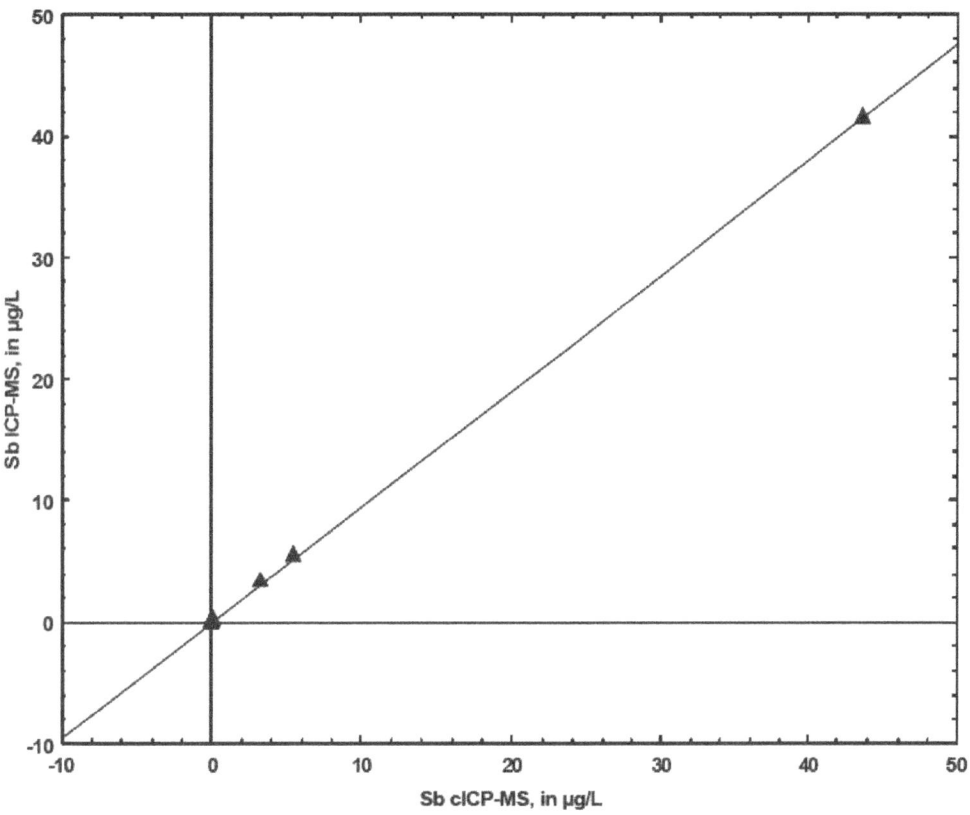

LINEAR REGRESSION EQUATION

Sb ICP-MS, in µg/L = .016 + .95 * Sb cICP-MS, in µg/L; R^2 = 1

Confidence Intervals
Sb ICP-MS, in µg/L vs. Sb cICP-MS, in µg/L

	Coefficient	95% Lower	95% Upper
Intercept	.016	.006	.026
Sb cICP-MS, in µg/L	.950	.949	.952

Figure A41. Linear regression analysis of antimony results from filtered water samples analyzed by inductively coupled plasma–mass spectrometry (ICP–MS) and collision/reaction cell inductively coupled plasma–mass spectrometry (cICP–MS). R^2 is the coefficient of determination. In the confidence intervals table, the Intercept and Sb cICP–MS coefficients are the *y*-intercept and slope, respectively. Results are in micrograms per liter (µg/L).

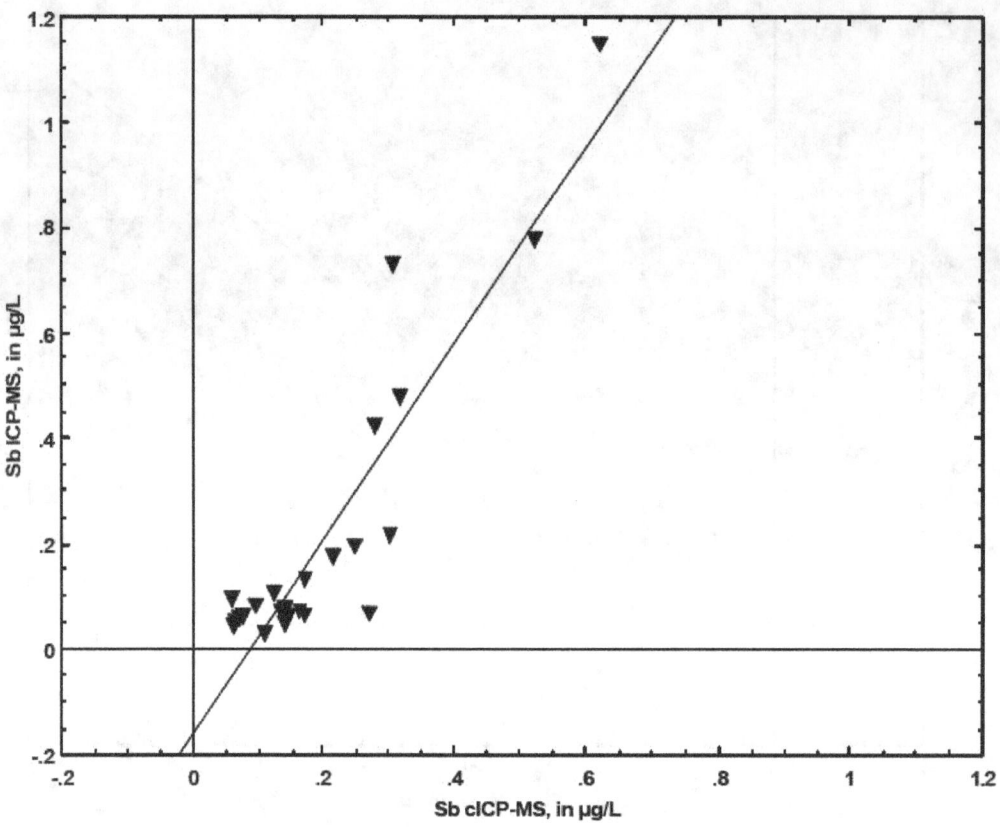

LINEAR REGRESSION EQUATION
Sb ICP-MS, in µg/L = -.161 + 1.846 * Sb cICP-MS, in µg/L; R^2 = .829

Confidence Intervals
Sb ICP-MS, in µg/L vs. Sb cICP-MS, in µg/L

	Coefficient	95% Lower	95% Upper
Intercept	-.161	-.249	-.073
Sb cICP-MS, in µg/L	1.846	1.484	2.207

Figure A42. Linear regression analysis of antimony results from whole-water recoverable digestates analyzed by inductively coupled plasma–mass spectrometry (ICP–MS) and collision/reaction cell inductively coupled plasma–mass spectrometry (cICP–MS). R^2 is the coefficient of determination. In the confidence intervals table, the Intercept and Sb cICP–MS coefficients are the *y*-intercept and slope, respectively. Results are in micrograms per liter (µg/L).

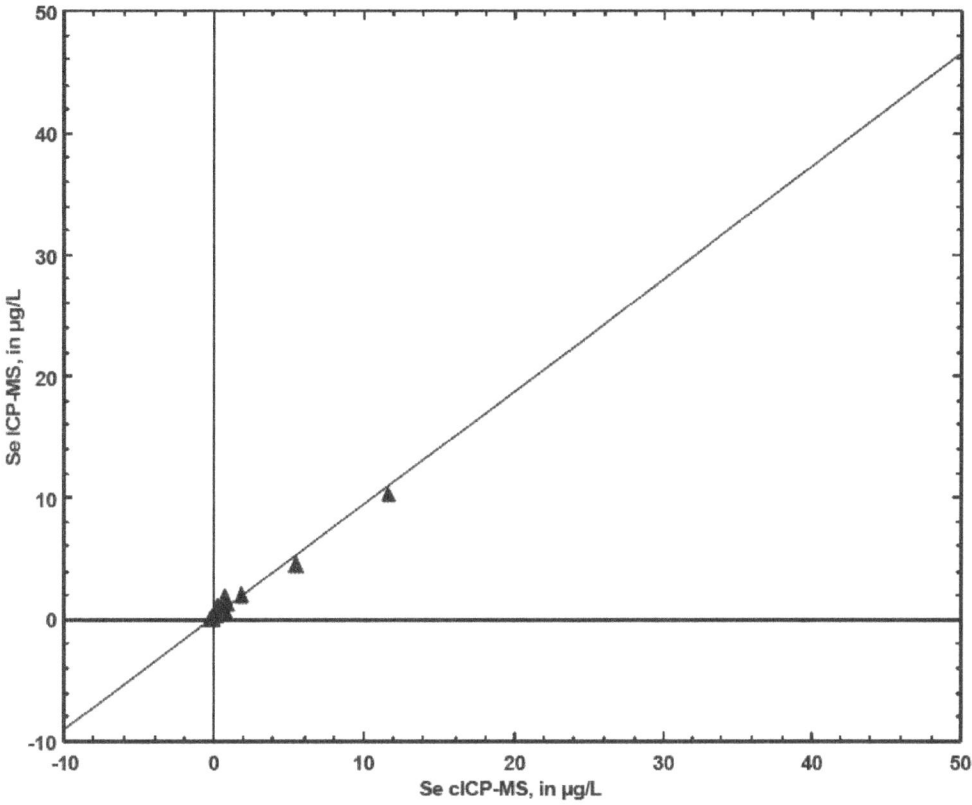

LINEAR REGRESSION EQUATION

Se ICP-MS, in µg/L = .134 + .929 * Se cICP-MS, in µg/L; R^2 = .998

Confidence Intervals
Se ICP-MS, in µg/L vs. Se cICP-MS, in µg/L

	Coefficient	95% Lower	95% Upper
Intercept	.134	-.035	.303
Se cICP-MS, in µg/L	.929	.913	.944

Figure A43. Linear regression analysis of selenium results from filtered water samples analyzed by inductively coupled plasma–mass spectrometry (ICP–MS) and collision/reaction cell inductively coupled plasma–mass spectrometry (cICP–MS). R^2 is the coefficient of determination. In the confidence intervals table, the Intercept and Se cICP–MS coefficients are the y-intercept and slope, respectively. Results are in micrograms per liter (µg/L).

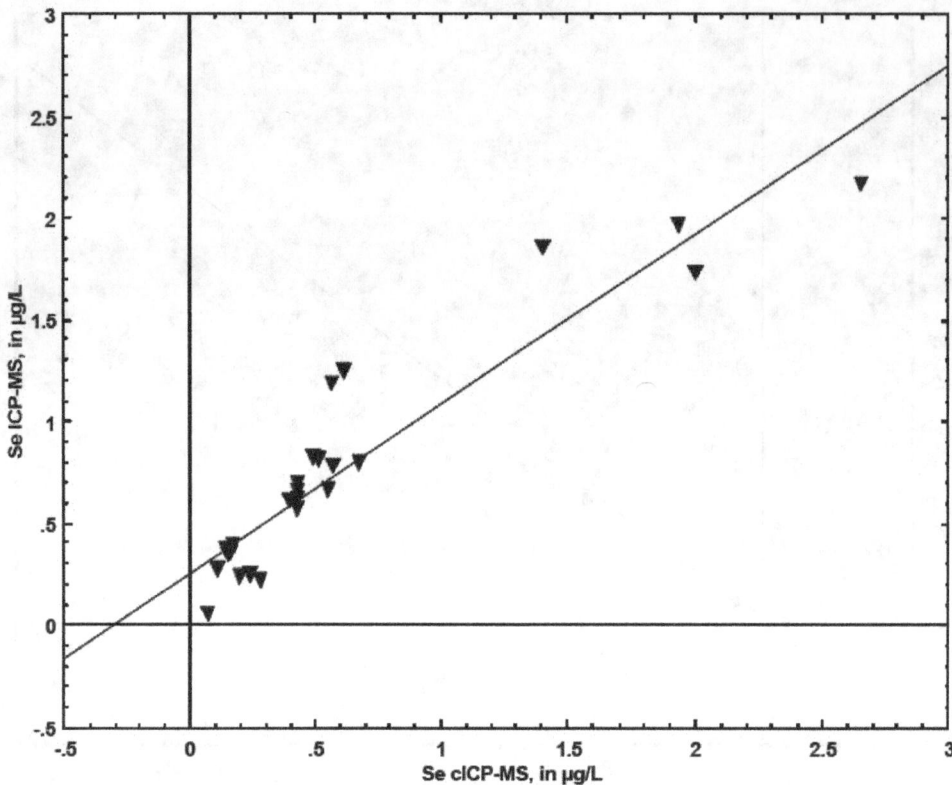

LINEAR REGRESSION EQUATION

Se ICP-MS, in µg/L = .255 + .831 * Se cICP-MS, in µg/L; R^2 = .875

Confidence Intervals
Se ICP-MS, in µg/L vs. Se cICP-MS, in µg/L

	Coefficient	95% Lower	95% Upper
Intercept	.255	.133	.378
Se cICP-MS, in µg/L	.831	.695	.967

Figure A44. Linear regression analysis of selenium results from whole-water recoverable digestates analyzed by inductively coupled plasma–mass spectrometry (ICP–MS) and collision/reaction cell inductively coupled plasma–mass spectrometry (cICP–MS). R^2 is the coefficient of determination. In the confidence intervals table, the Intercept and Se cICP–MS coefficients are the y-intercept and slope, respectively. Results are in micrograms per liter (µg/L).

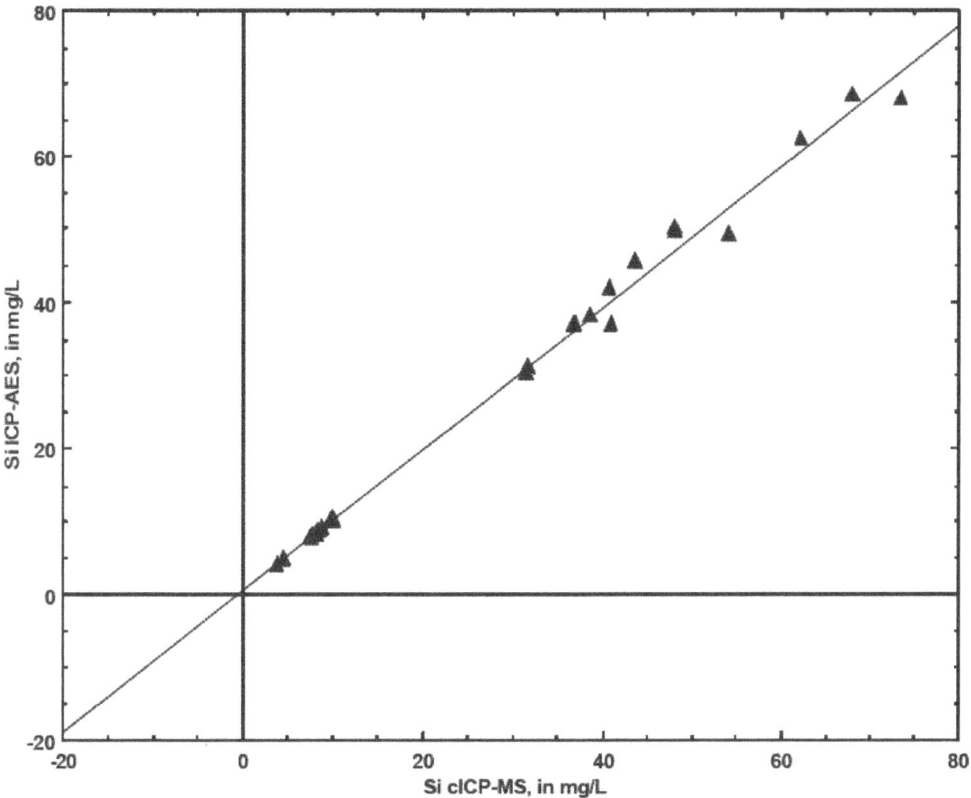

LINEAR REGRESSION EQUATION

Si ICP-AES, in mg/L = .339 + .971 * Si cICP-MS, in mg/L; R^2 = .993

Confidence Intervals
Si ICP-AES, in mg/L vs. Si cICP-MS, in mg/L

	Coefficient	95% Lower	95% Upper
Intercept	.339	-.890	1.567
Si cICP-MS, in mg/L	.971	.937	1.005

Figure A45. Linear regression analysis of silicon (as SiO_2) results from filtered water samples analyzed by inductively coupled plasma–atomic emission spectrometry (ICP–AES) and collision/reaction cell inductively coupled plasma–mass spectrometry (cICP–MS). R^2 is the coefficient of determination. In the confidence intervals table, the Intercept and Si cICP–MS coefficients are the y-intercept and slope, respectively. Results are in milligrams per liter (mg/L).

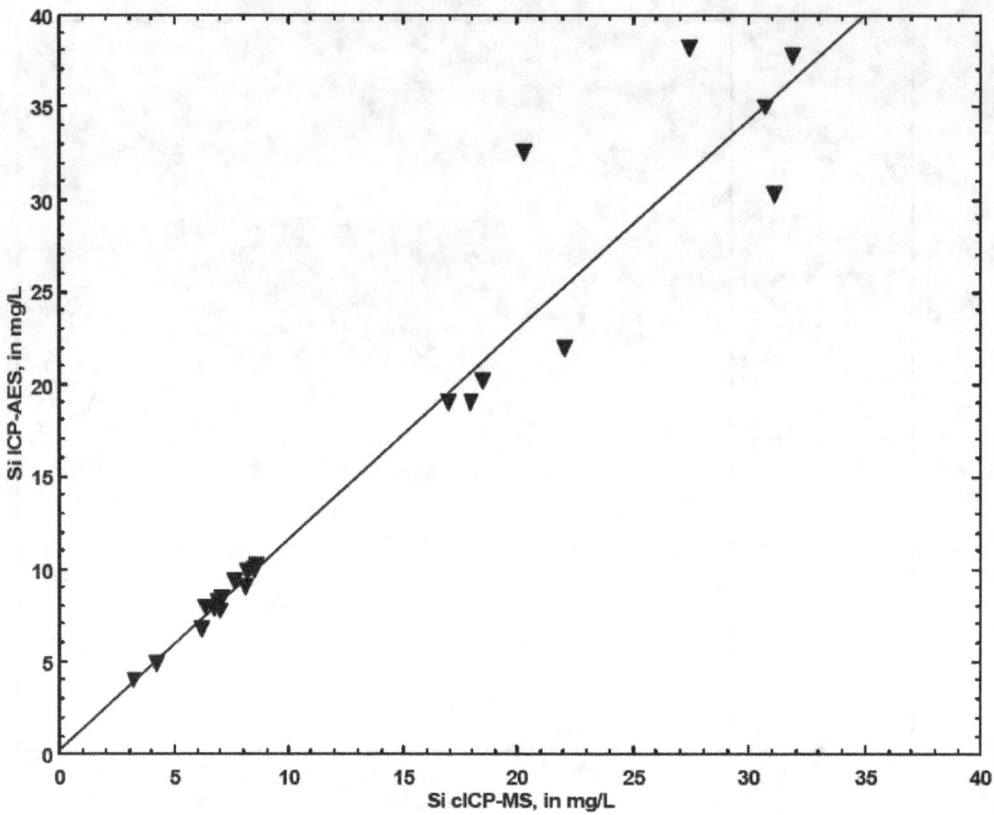

LINEAR REGRESSION EQUATION

Si ICP-AES, in mg/L = .238 + 1.136 * Si cICP-MS, in mg/L; R^2 = .946

Confidence Intervals
Si ICP-AES, in mg/L vs. Si cICP-MS, in mg/L

	Coefficient	95% Lower	95% Upper
Intercept	.238	-1.999	2.475
Si cICP-MS, in mg/L	1.136	1.014	1.259

Figure A46. Linear regression analysis of silicon (as SiO_2) results from whole-water recoverable digestates analyzed by inductively coupled plasma–atomic emission spectrometry (ICP–AES) and collision/reaction cell inductively coupled plasma–mass spectrometry (cICP–MS). R^2 is the coefficient of determination. In the confidence intervals table, the Intercept and Si cICP–MS coefficients are the *y*-intercept and slope, respectively. Results are in milligrams per liter (mg/L).

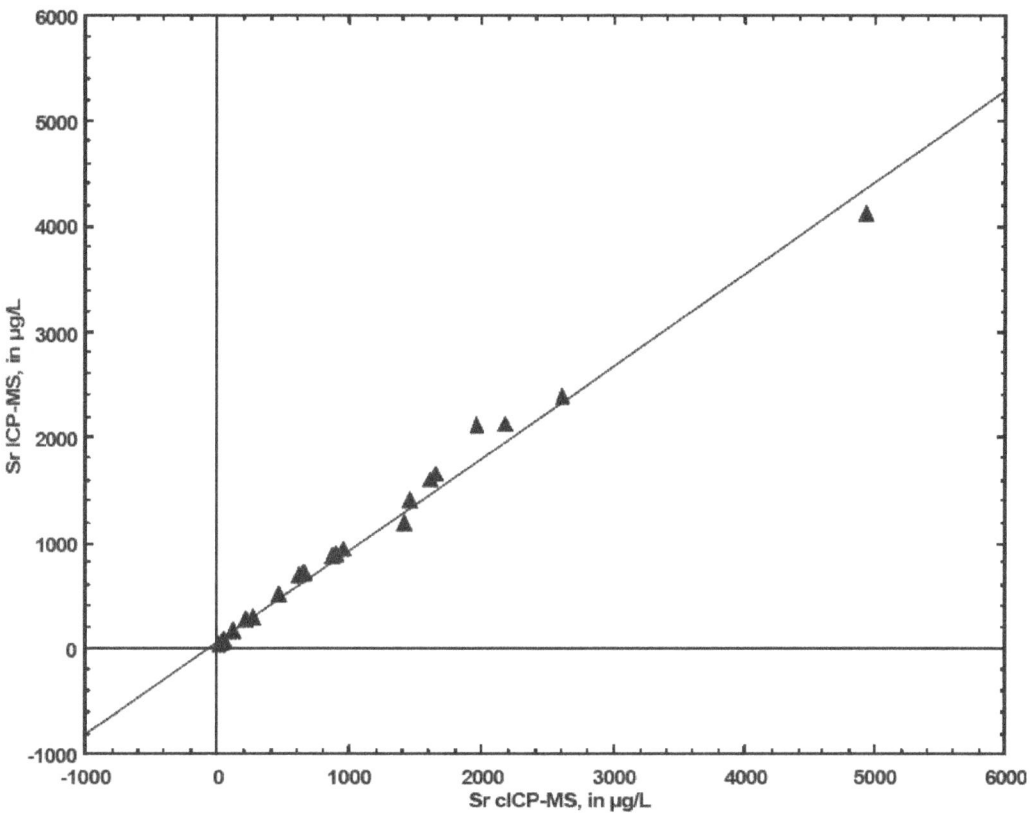

LINEAR REGRESSION EQUATION

Sr ICP-MS, in µg/L = 68.799 + .87 * Sr cICP-MS, in µg/L; R^2 = .988

Confidence Intervals
Sr ICP-MS, in µg/L vs. Sr cICP-MS, in µg/L

	Coefficient	95% Lower	95% Upper
Intercept	68.799	9.040	128.559
Sr cICP-MS, in µg/L	.870	.829	.911

Figure A47. Linear regression analysis of strontium results from filtered water samples analyzed by inductively coupled plasma–mass spectrometry (ICP–MS) and collision/reaction cell inductively coupled plasma–mass spectrometry (cICP–MS). R^2 is the coefficient of determination. In the confidence intervals table, the Intercept and Sr cICP–MS coefficients are the y-intercept and slope, respectively. Results are in micrograms per liter (µg/L).

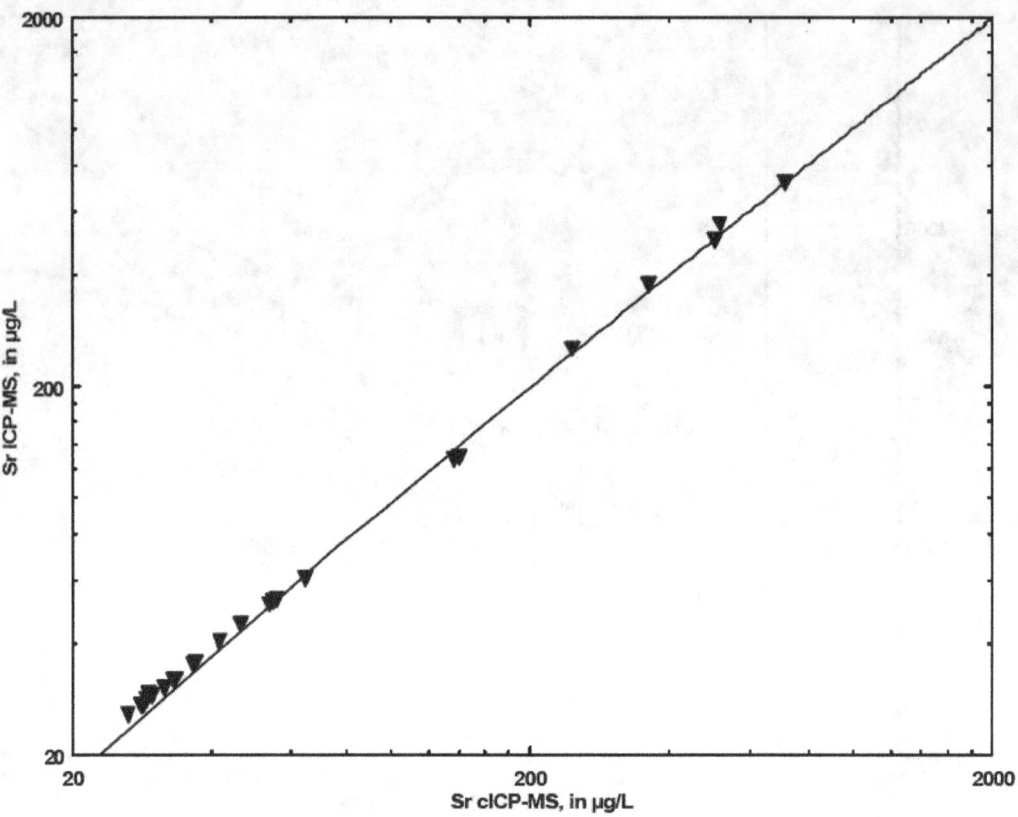

LINEAR REGRESSION EQUATION

Sr ICP-MS, in µg/L = -2.933 + 1.004 * Sr cICP-MS, in µg/L; R^2 = .999

Confidence Intervals
Sr ICP-MS, in µg/L vs. Sr cICP-MS, in µg/L

	Coefficient	95% Lower	95% Upper
Intercept	-2.933	-6.698	.831
Sr cICP-MS, in µg/L	1.004	.987	1.020

Figure A48. Linear regression analysis of strontium results from whole-water recoverable digestates analyzed by inductively coupled plasma–mass spectrometry (ICP–MS) and collision/reaction cell inductively coupled plasma–mass spectrometry (cICP–MS). R^2 is the coefficient of determination. In the confidence intervals table, the Intercept and Sr cICP–MS coefficients are the *y*-intercept and slope, respectively. Results are in micrograms per liter (µg/L).

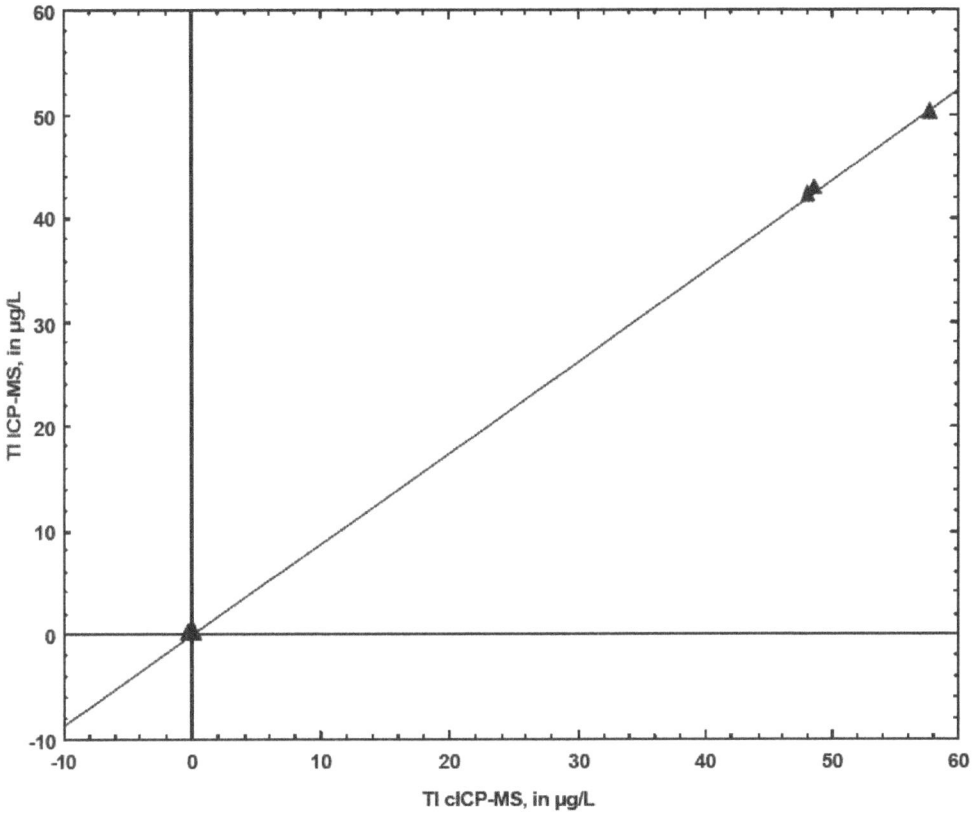

LINEAR REGRESSION EQUATION

Tl ICP-MS, in µg/L = .011 + .873 * Tl cICP-MS, in µg/L; R^2 = 1

Confidence Intervals
Tl ICP-MS, in µg/L vs. Tl cICP-MS, in µg/L

	Coefficient	95% Lower	95% Upper
Intercept	.011	-.035	.057
Tl cICP-MS, in µg/L	.873	.870	.875

Figure A49. Linear regression analysis of thallium results from filtered water samples analyzed by inductively coupled plasma–mass spectrometry (ICP–MS) and collision/reaction cell inductively coupled plasma–mass spectrometry (cICP–MS). R^2 is the coefficient of determination. In the confidence intervals table, the Intercept and Tl cICP–MS coefficients are the y-intercept and slope, respectively. Results are in micrograms per liter (µg/L).

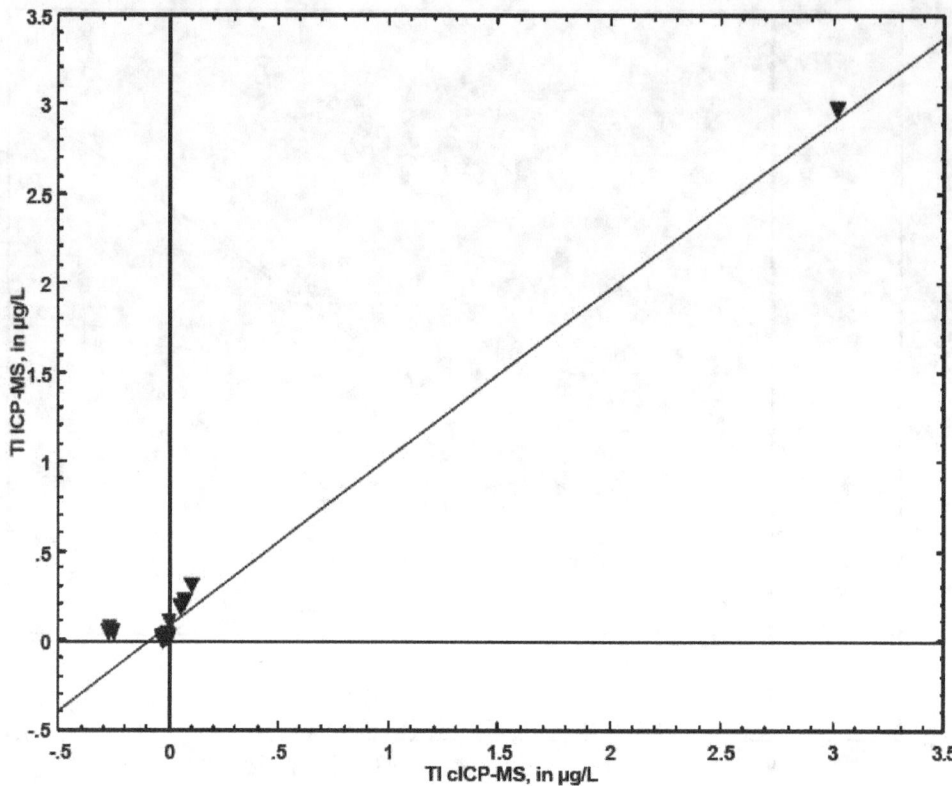

LINEAR REGRESSION EQUATION

Tl ICP-MS, in µg/L = .072 + .942 * Tl cICP-MS, in µg/L; R^2 = .973

Confidence Intervals
Tl ICP-MS, in µg/L vs. Tl cICP-MS, in µg/L

	Coefficient	95% Lower	95% Upper
Intercept	.072	.031	.113
Tl cICP-MS, in µg/L	.942	.874	1.009

Figure A50. Linear regression analysis of thallium results from whole-water recoverable digestates analyzed by inductively coupled plasma–mass spectrometry (ICP–MS) and collision/reaction cell inductively coupled plasma–mass spectrometry (cICP–MS). R^2 is the coefficient of determination. In the confidence intervals table, the Intercept and Tl cICP–MS coefficients are the y-intercept and slope, respectively. Results are in micrograms per liter (µg/L).

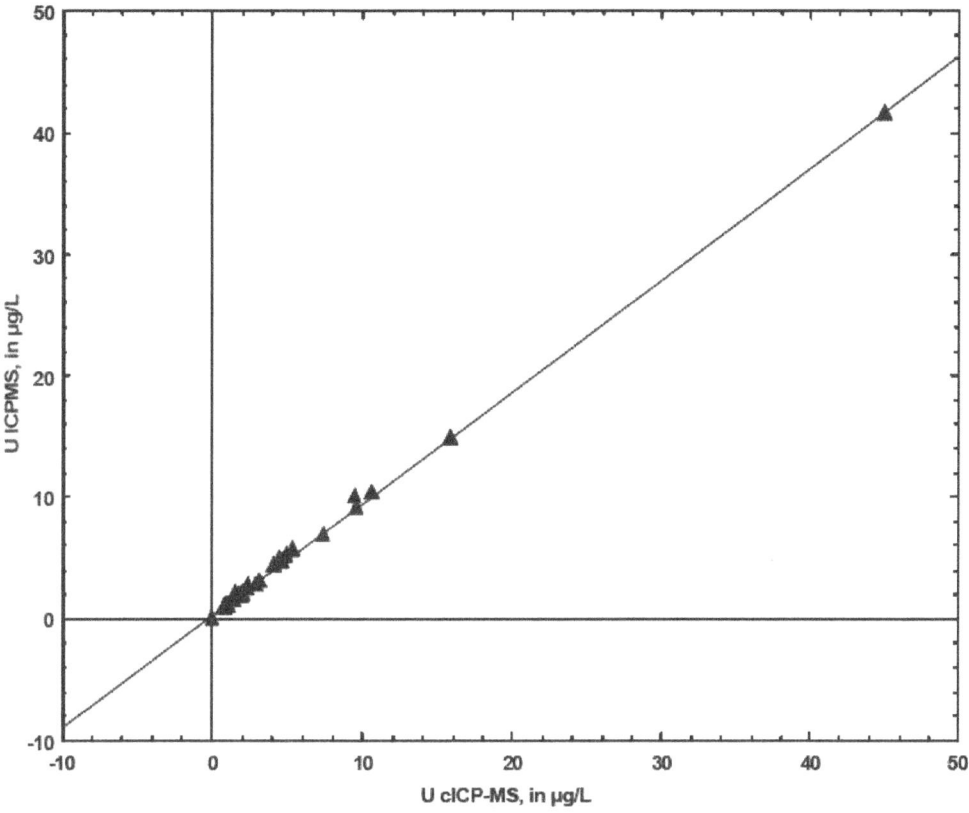

LINEAR REGRESSION EQUATION

U ICPMS, in µg/L = .246 + .92 * U cICP-MS, in µg/L; R^2 = .999

Confidence Intervals
U ICPMS, in µg/L vs. U cICP-MS, in µg/L

	Coefficient	95% Lower	95% Upper
Intercept	.246	.089	.403
U cICP-MS, in µg/L	.920	.905	.935

Figure A51. Linear regression analysis of uranium results from filtered water samples analyzed by inductively coupled plasma–mass spectrometry (ICP–MS) and collision/reaction cell inductively coupled plasma–mass spectrometry (cICP–MS). R^2 is the coefficient of determination. In the confidence intervals table, the Intercept and U cICP–MS coefficients are the *y*-intercept and slope, respectively. Results are in micrograms per liter (µg/L).

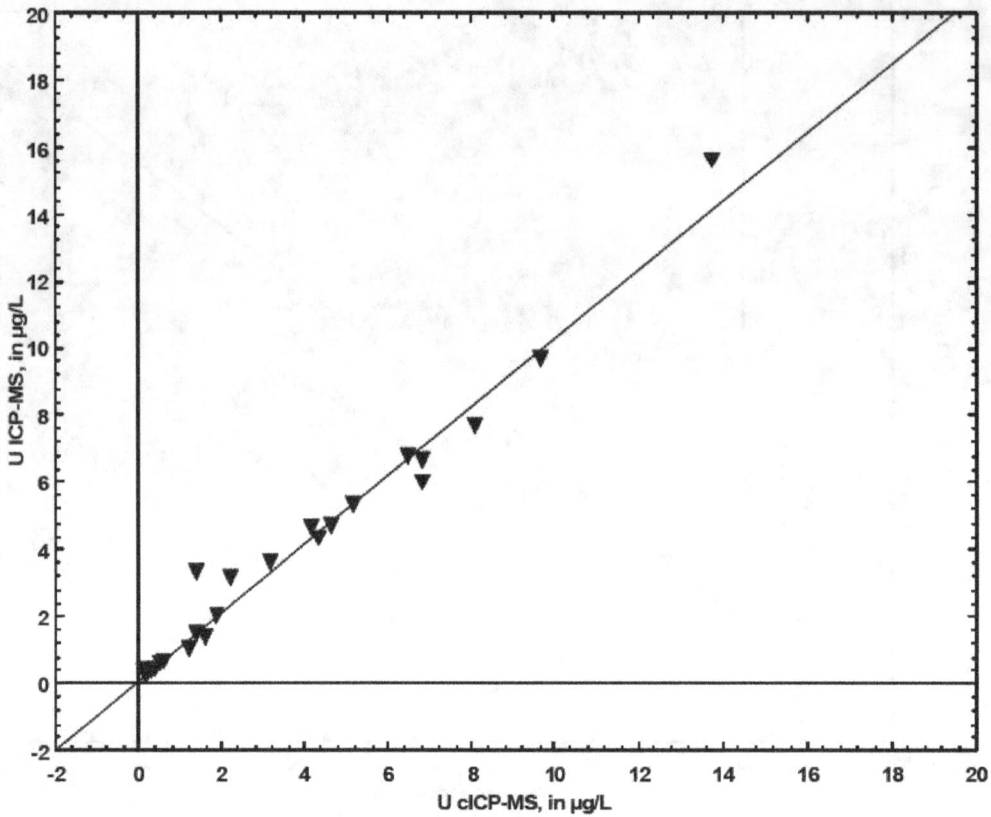

LINEAR REGRESSION EQUATION
U ICP-MS, in µg/L = .046 + 1.024 * U cICP-MS, in µg/L; R^2 = .973

Confidence Intervals
U ICP-MS, in µg/L vs. U cICP-MS, in µg/L

	Coefficient	95% Lower	95% Upper
Intercept	.046	-.313	.405
U cICP-MS, in µg/L	1.024	.951	1.097

Figure A52. Linear regression analysis of uranium results from whole-water recoverable digestates analyzed by inductively coupled plasma–mass spectrometry (ICP–MS) and collision/reaction cell inductively coupled plasma–mass spectrometry (cICP–MS). R^2 is the coefficient of determination. In the confidence intervals table, the Intercept and U cICP–MS coefficients are the y-intercept and slope, respectively. Results are in micrograms per liter (µg/L).

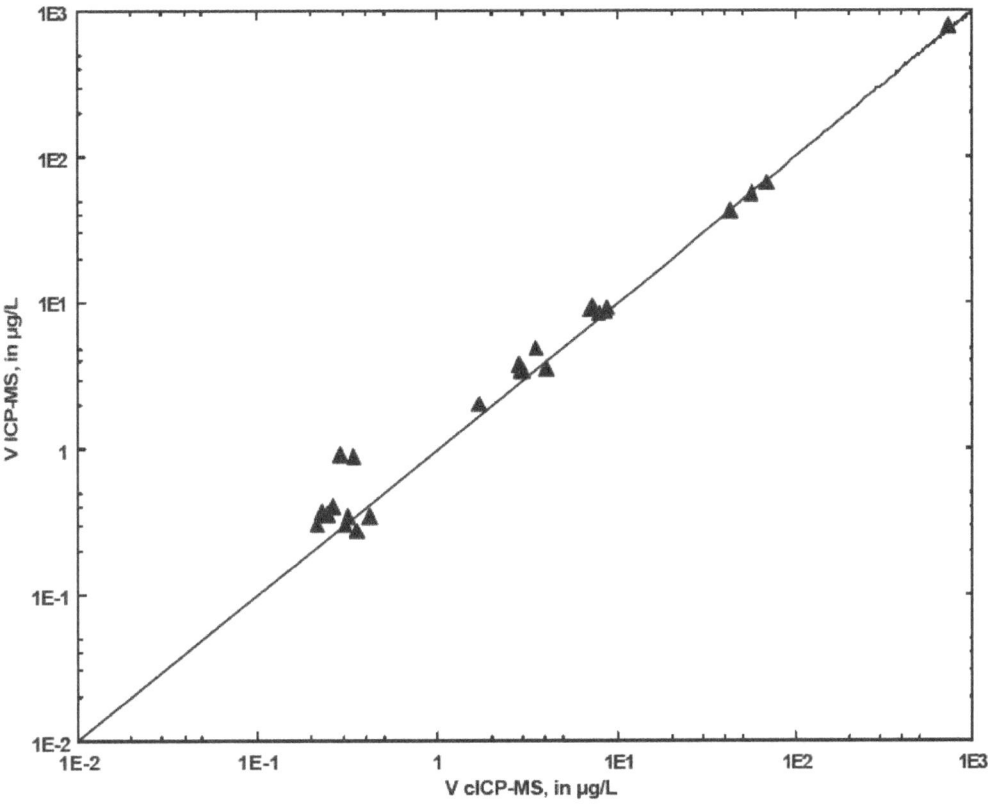

LINEAR REGRESSION EQUATION

V ICP-MS, in µg/L = -.162 + 1.01 * V cICP-MS, in µg/L; R^2 = 1

Confidence Intervals
V ICP-MS, in µg/L vs. V cICP-MS, in µg/L

	Coefficient	95% Lower	95% Upper
Intercept	-.162	-.803	.478
V cICP-MS, in µg/L	1.010	1.005	1.014

Figure A53. Linear regression analysis of vanadium results from filtered water samples analyzed by inductively coupled plasma–mass spectrometry (ICP–MS) and collision/reaction cell inductively coupled plasma–mass spectrometry (cICP–MS). R^2 is the coefficient of determination. In the confidence intervals table, the Intercept and V cICP–MS coefficients are the y-intercept and slope, respectively. Results are in micrograms per liter (µg/L).

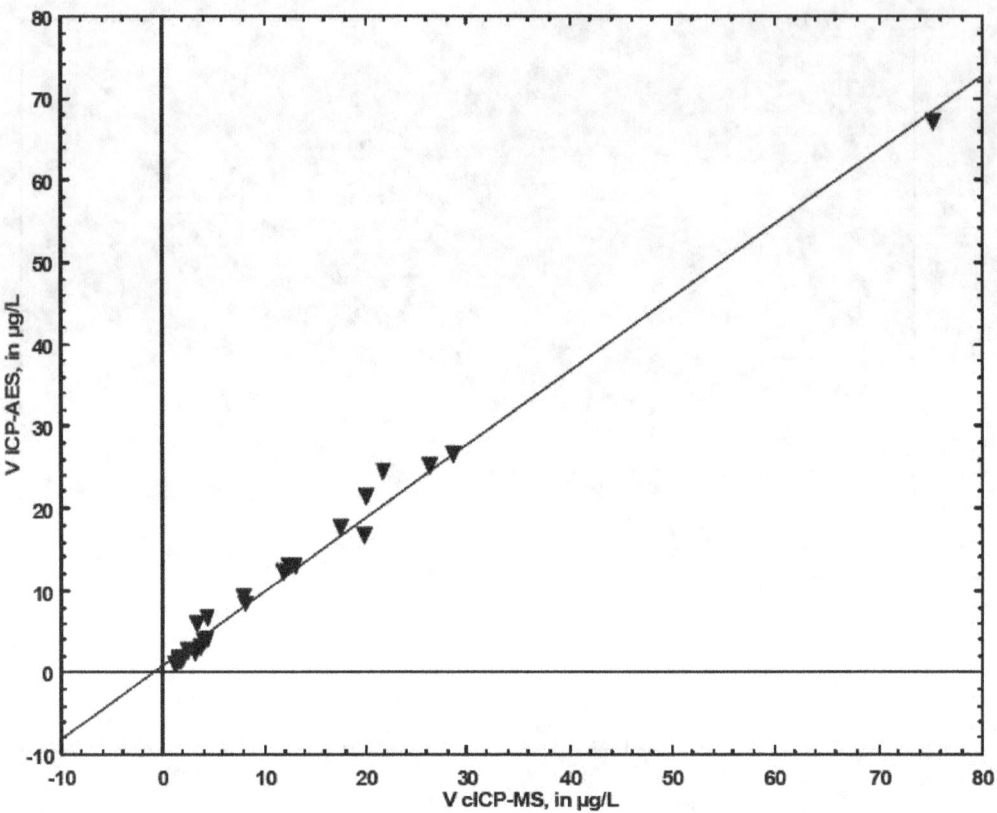

LINEAR REGRESSION EQUATION

V ICP-AES, in µg/L = .716 + .9 * V cICP-MS, in µg/L; R^2 = .99

Confidence Intervals
V ICP-AES, in µg/L vs. V cICP-MS, in µg/L

	Coefficient	95% Lower	95% Upper
Intercept	.716	-.035	1.466
V cICP-MS, in µg/L	.900	.862	.939

Figure A54. Linear regression analysis of vanadium results from whole-water recoverable digestates analyzed by inductively coupled plasma–atomic emission spectrometry (ICP–AES) and collision/reaction cell inductively coupled plasma–mass spectrometry (cICP–MS). R^2 is the coefficient of determination. In the confidence intervals table, the Intercept and V cICP–MS coefficients are the *y*-intercept and slope, respectively. Results are in micrograms per liter (µg/L).

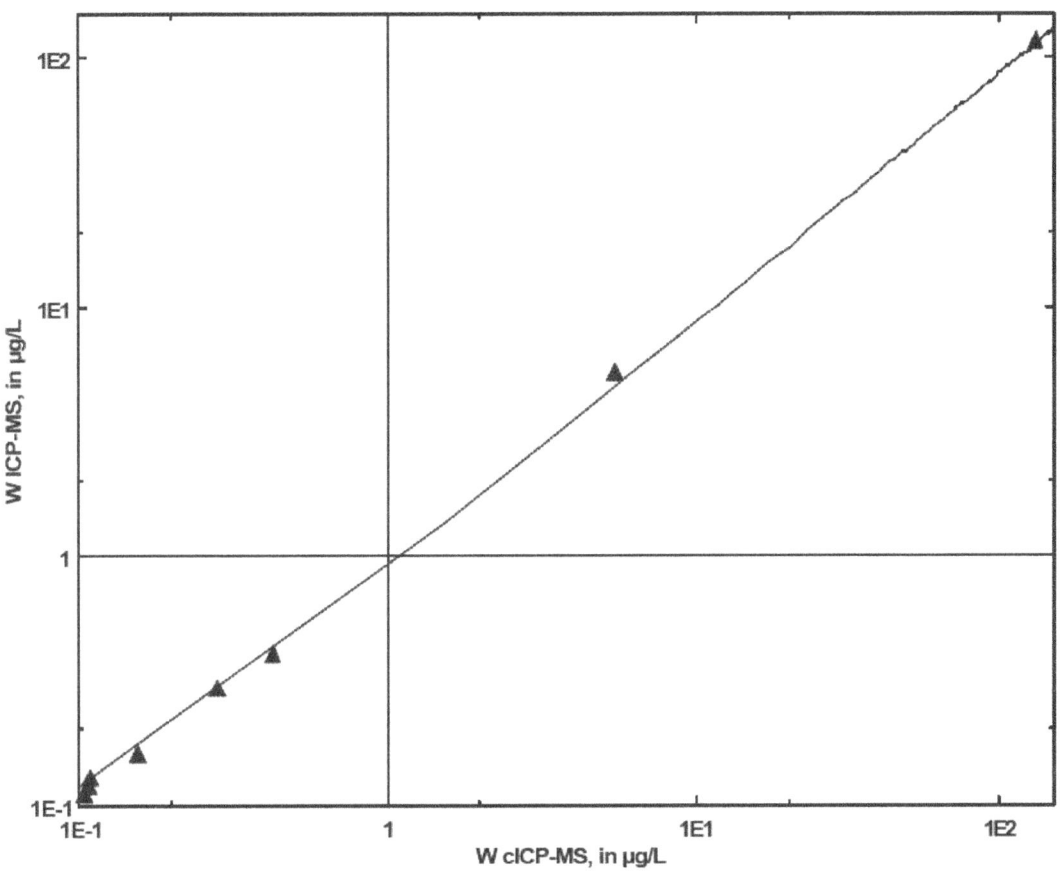

LINEAR REGRESSION EQUATION

W ICP-MS, in µg/L = .031 + .874 * W cICP-MS, in µg/L; R^2 = 1

Confidence Intervals
W ICP-MS, in µg/L vs. W cICP-MS, in µg/L

	Coefficient	95% Lower	95% Upper
Intercept	.031	-.013	.076
W cICP-MS, in µg/L	.874	.873	.876

Figure A55. Linear regression analysis of tungsten results from filtered water samples analyzed by inductively coupled plasma–mass spectrometry (ICP–MS) and collision/reaction cell inductively coupled plasma–mass spectrometry (cICP–MS). R^2 is the coefficient of determination. In the confidence intervals table, the Intercept and W cICP–MS coefficients are the *y*-intercept and slope, respectively. Results are in micrograms per liter (µg/L).

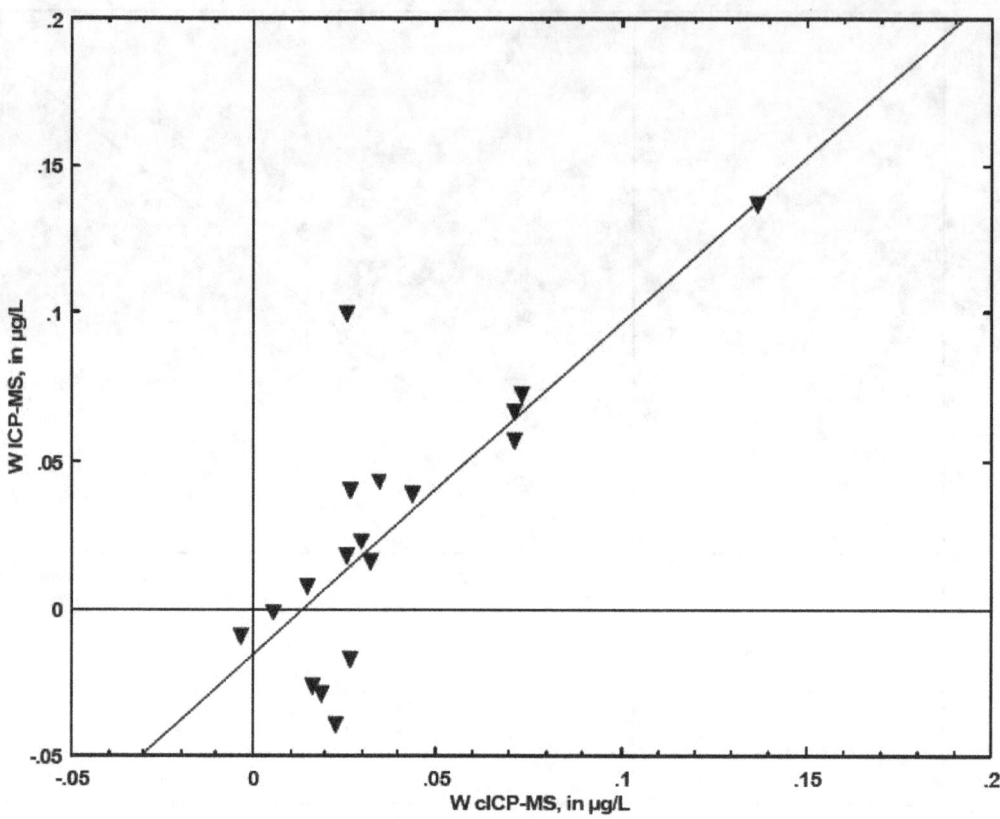

LINEAR REGRESSION EQUATION

W ICP-MS, in µg/L = -.016 + 1.122 * W cICP-MS, in µg/L; R^2 = .617

Confidence Intervals
W ICP-MS, in µg/L vs. W cICP-MS, in µg/L

	Coefficient	95% Lower	95% Upper
Intercept	-.016	-.039	.008
W cICP-MS, in µg/L	1.122	.654	1.589

Figure A56. Linear regression analysis of tungsten results from whole-water recoverable digestates analyzed by inductively coupled plasma–mass spectrometry (ICP–MS) and collision/reaction cell inductively coupled plasma–mass spectrometry (cICP–MS). R^2 is the coefficient of determination. In the confidence intervals table, the Intercept and W cICP–MS coefficients are the *y*-intercept and slope, respectively. Results are in micrograms per liter (µg/L).

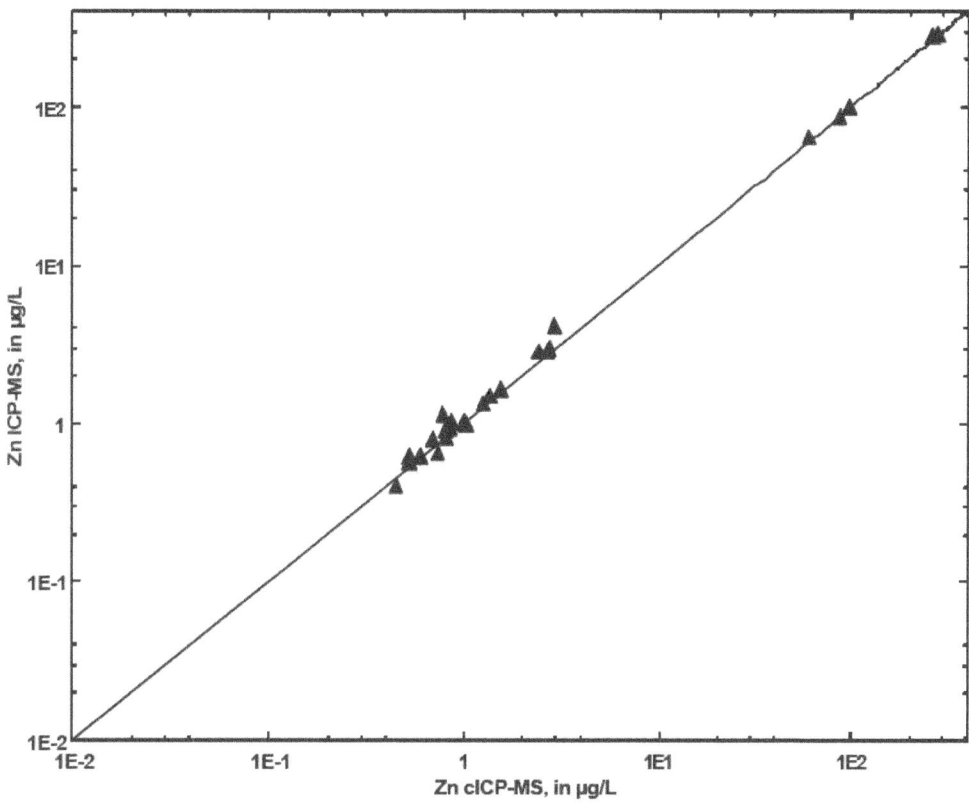

LINEAR REGRESSION EQUATION

Zn ICP-MS, in µg/L = -.052 + 1.02 * Zn cICP-MS, in µg/L; R^2 = 1

Confidence Intervals
Zn ICP-MS, in µg/L vs. Zn cICP-MS, in µg/L

	Coefficient	95% Lower	95% Upper
Intercept	-.052	-.692	.587
Zn cICP-MS, in µg/L	1.020	1.013	1.028

Figure A57. Linear regression analysis of zinc results from filtered water samples analyzed by inductively coupled plasma–mass spectrometry (ICP–MS) and collision/reaction cell inductively coupled plasma–mass spectrometry (cICP–MS). R^2 is the coefficient of determination. In the confidence intervals table, the Intercept and Zn cICP–MS coefficients are the *y*-intercept and slope, respectively. Results are in micrograms per liter (µg/L).

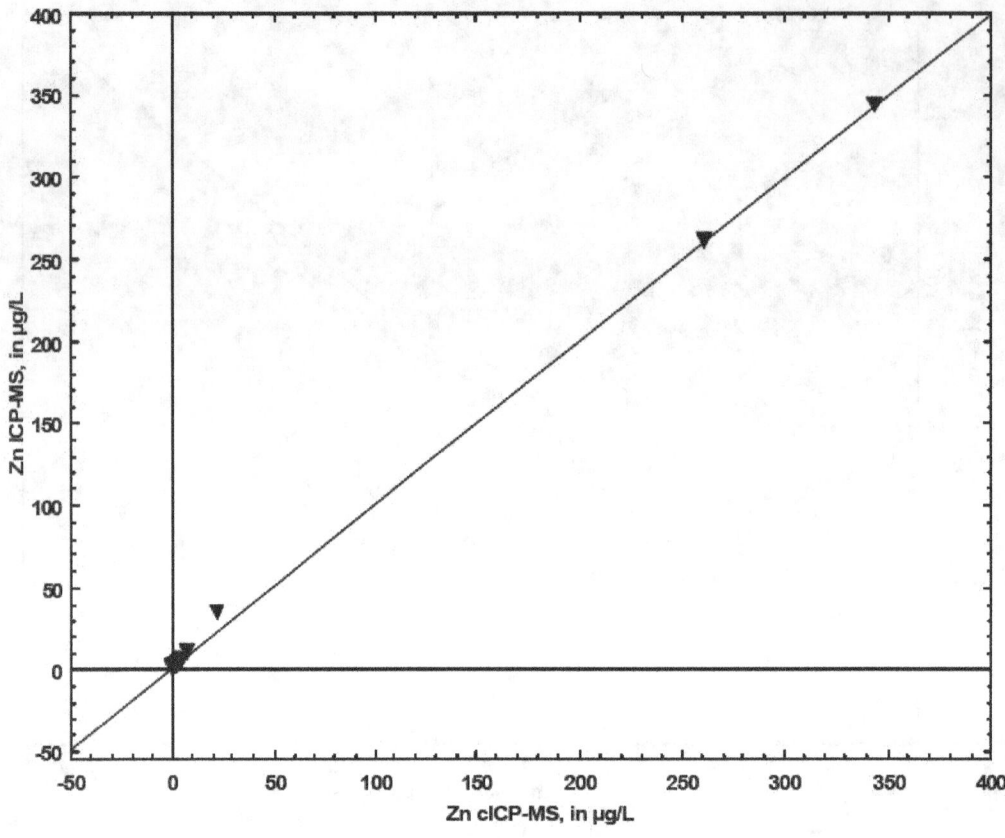

LINEAR REGRESSION EQUATION

Zn ICP-MS, in µg/L = .538 + .995 * Zn cICP-MS, in µg/L; R^2 = .999

Confidence Intervals
Zn ICP-MS, in µg/L vs. Zn cICP-MS, in µg/L

	Coefficient	95% Lower	95% Upper
Intercept	.538	-.560	1.637
Zn cICP-MS, in µg/L	.995	.983	1.008

Figure A58. Linear regression analysis of zinc results from whole-water recoverable digestates analyzed by inductively coupled plasma–mass spectrometry (ICP–MS) and collision/reaction cell inductively coupled plasma–mass spectrometry (cICP–MS). R^2 is the coefficient of determination. In the confidence intervals table, the Intercept and Zn cICP–MS coefficients are the *y*-intercept and slope, respectively. Results are in micrograms per liter (µg/L).

Glossary

Bias Systematic error that is manifested as a consistent positive or negative deviation from the known or true value (Taylor, 1990). The error can be positive or negative. Positive error can result from sample contamination or spectral interference, whereas negative error can result from analyte loss or signal suppression. Bias differs from random error which shows no such consistent or systematic deviation.

F-pseudosigma A nonparametric, resistant measure of data spread defined as the inter-quartile range of the data divided by 1.349 (Hoaglin and others, 1983). F-pseudosigma and the standard deviation of the data will be nearly equivalent if the data have a near-normal distribution.

Long-term method detection level (LT–MDL) The minimum concentration of an element that can be measured and reported with 99-percent confidence that the elemental concentration is greater than zero. The LT–MDL is calculated from replicate analyses of samples fortified with all the method elements, and includes variability introduced by multiple instruments, multiple analysts, and multiple calibrations over 6 to 12 months (Childress and others, 1999).

Matrix The substrate of a test sample (National Environmental Laboratory Accreditation Conference, 2003).

Matrix spike A sample prepared by adding a known mass of target analyte to a specified amount of matrix sample for which an independent estimate of target analyte concentration is available. Matrix spike is used to determine the effect of the matrix on a method's recovery efficiency (National Environmental Laboratory Accreditation Conference, 2003).

Method detection limit (MDL) The minimum concentration of an element that can be measured and reported with 99-percent confidence that the elemental concentration is greater than zero and is determined from analysis of a sample in a given matrix containing the element of interest (U.S. Environmental Protection Agency, 2000).

Spike A known mass of target analyte added to a blank sample or subsample; used to determine the recovery efficiency or for other quality-control purposes (National Environmental Laboratory Accreditation Conference, 2003).

Variability Random error in independent measurements as the result of repeated application of the process under specific conditions. Precision is a measure of variability in experimental data that can be statistically characterized by the standard deviation (Taylor, 1990).

www.ingramcontent.com/pod-product-compliance
Lightning Source LLC
Chambersburg PA
CBHW081547170526

45166CB00009B/2610